P9-CAV-574

Plain Talk About The Weather

Everyone talks about the weather—farmer and executive, housewife and gardener, yachtsman and sports fan. But few people understand how to analyze the weather and do something about it for their own benefit.

Here's a handy, concise and highly readable, illustrated explanation of the processes which make up the weather, that gives you the basic facts you need to be your own forecaster. With the help of this book, and the inexpensive equipment it recommends, you can learn how to predict the weather and intelligently control its ill-effects on your daily affairs.

Among the fascinating facts in this entertaining book you will find information about the layers of air above our world, what makes temperatures rise or fall, how to identify and understand the significance of various clouds, what causes storms, and how to read and interpret the weather maps in your daily newspaper.

Robert Moore Fischer, who is presently working toward his Ph.D. at Columbia University, is a First Lieutenant, Weather Officer, in the United States Air Force Reserve. The original edition of this book was published by Harper & Brothers.

How to Know and Predict the Weather

(Original title: How About the Weather)

by

ROBERT MOORE FISHER

With a Foreword by

ERNEST J. CHRISTIE

Meteorologist in Charge

United States Weather Bureau Office

New York, N. Y.

A MENTOR BOOK

Published by THE NEW AMERICAN LIBRARY

Published as a MENTOR BOOK
By Arrangement with Harper & Brothers

FIRST PRINTING, APRIL, 1953

MENTOR BOOKS are published by
The New American Library of World Literature, Inc.
501 Madison Avenue, New York 22, New York

PRINTED IN THE UNITED STATES OF AMERICA

TO MY PARENTS
without whom this book would
have been impossible

Foreword

EVERYBODY talks about the weather. It is one of the chief topics of conversation among friends whenever they meet at lunch, in an elevator, or on a bus. Moreover, everybody reads about the weather, for it is often an outstanding news item of the day, and may crowd war-talk or other "big news" stories off the front pages of local newspapers. It is easy to understand why weather is so important to friends and newspapers, because weather in one or another of its forms influences everyone. For some people, it may be just a little more pain in a rheumatic joint; for others, it may be a dollar-and-cents proposition affecting the returns of a business. So be you banker, baker, or candlestick maker, the weather plays an important role in what you do and how you do it.

A former Chief of the United States Weather Bureau once said that weather and climate make up one of our national resources. This is indeed the truth; in fact, we need not stretch our imaginations far to discover that weather and climate are one of our most vital natural resources, for so many other natural resources depend upon weather and climate for their existence—the brooks, rivers, and lakes; the woods and forests; yes, even the minerals in the soil.

However, unlike most natural resources, which can be worked on and guided by man's efforts, weather and climate defy manipulation. No matter how hard we try, the physical processes which produce weather and climate will never be chained by slide rules or steam shovels.

Yet all is not hopeless. In the past, we have learned to use other natural resources to advantage by studying their behavior and then by setting up a program by means of which we could get the most out of them. If we follow the same procedure in dealing with weather and climate, we can plan our activities so as to limit the amount of damage wrought in times of storm, or so as to maximize the benefits accruing from the day-to-day weather phenomena or from the changes therein.

In the days of old, farmers and fishermen became keen weather prophets because they recognized the fact that it was to their advantage to know what was brewing for the morrow. Today, nearly everyone acknowledges this fact and tries his hand at "guessing" what tomorrow's weather will be. Some of these guesses are based on fiction, but more and more of them are beginning to be based on fact.

Businessmen, for example, are beginning to utilize the services of Weather Bureau forecasts and advisories to an increasing extent, for they realize that the weather affects many phases of their operations from production to sales. By analyzing the problem of weather in relation to business, they have placed the weather on a cost basis which not only makes dollars, but also sense. Furthermore, they have learned something about meteorology itself, so that they are able to understand some of the problems of the weather forecasters with whom they talk every day. And when they have gotten the desired information about the trend of tomorrow's weather, businessmen have been able to make significant decisions as to what procedures will be most likely to benefit their businesses.

This type of analysis, which is far removed from the by-guess-and-by-gosh techniques of past centuries, is both businesslike and intelligent. Whether you are in business or out of doors, whether you work in the kitchen or on the railroad, you can apply the same sort of analysis to your daily affairs and do something about the weather for your own benefit.

This book will help you get started in the right direction—which is up. It will acquaint you with the processes which make up the weather. It will show you the meaning of certain weather signs, and it will give you some hints about how to interpret the weather maps you see each day in your local newspapers. The author has written the book in an appealing style, which makes it easy, pleasant, and profitable to read, although it contains loads of technical information. He writes about the entire weather picture in a way that makes even a discussion about the pressure of the atmosphere an interesting reading experience. What you read in this book about the weather—and what you learn later by following it—will undoubtedly be a help to you in your future dealings with one of the most common of all topics of conversation, "The Weather."

ERNEST J. CHRISTIE
Meteorologist in Charge
United States Weather Bureau Office
New York, N. Y.
1951

Preface

HERE is a book about the weather for the layman. As such, perhaps you're praying that the weather will be fair tomorrow so you can fly to Chicago, hang out the wash, or go golfing or sailing. Perhaps you're wondering just why the weather will be fair, or what will cause it to be wet in spite of your plans. Perhaps you're wishing that you could become better acquainted with the weather, and be able to anticipate its major changes by forecasting them yourself. I hope that this book will answer some of your questions as well as stimulate your interest in the world that lies overhead.

This volume will be of use to the Weather Bureau forecaster and to the airplane pilot and hostess, all of whom are concerned with explaining the weather to the layman.

The book will also enable high-school and college students to refresh their knowledge of the atmosphere and its habits.

But above all, it will help you—the layman—to get an informal picture of the clouds, wind, rain, fog, and every other weather element which influences your daily life. It will likewise help you to interpret the weather maps published in the newspapers, so that you can forecast tomorrow's weather today. In effect, *How to Know and Predict the Weather* will make you weather-wise instead of otherwise.

Since it is a book for the layman, complicated terms and explanations have been omitted whenever possible. Furthermore, many of the numerical values included are approximate. Technical precision must not be expected on every page. For that item, you must turn to a textbook, where it properly belongs.

The book grew out of the experiences I had while forecasting the weather in the Army Air Force during World War II. Somewhat to my surprise, I found that everybody was interested in the weather. Not only was the weather a common subject of conversation, but it was also a matter of curiosity. There were some people, of course, who could distinguish a sprawling stratus from a cottony cumulus cloud. But none of them knew how those clouds are manufactured or what significance they hold for tomorrow's weather. Yet they all were inquisitive.

Some were curious only to the extent of asking, "Hey, how about the weather? Will it rain tomorrow, or can I go to the ball game?" The reply was not always to go to the ball game.

Others were more intrigued. They wanted to learn about the history of meteorology and about its development as one of the youngest of our sciences. They wanted to know why the barometer was high in good weather and low in bad. What causes my permanent wave to change its appearance in rainy weather? Why is the southwest of the United States a desert, while the southeast is a park? What's a front, and who's in front of what? All of their questions, moreover, indicated a common desire to understand and forecast the weather so they could outguess the weatherman themselves.

In response to their interest and aspirations, this book is presented. Most textbooks are too advanced and technically worded to do the job. On the other hand, many popular books are written for someone who likes to discuss *moist adiabatic lapse rates* or *cyclonic vorticity*. In addition, these books fail to give the average reader an idea of how to forecast the weather by interpreting newspaper weather maps, as well as by watching the sky and wind, and possibly by reading the barometer and thermometer. I trust that this book will evoke not only an appreciation of the vast intricacy of the atmosphere, but also some sympathy for the professional meteorologists who fathom its complexity with remarkable results.

The discussion is limited to the weather within the United States.

All temperature values are given in degrees Fahrenheit.

All cloud elevations, unless otherwise indicated, are given as feet above the ground.

Finally, I wish to acknowledge my gratitude to the

following people and firms for their generous assistance: the New York and Washington offices of the U.S. Weather Bureau, Department of Commerce, for cloud photographs and helpful advice; the Public Relations Department, New York office, of the Standard Oil Company (N.J.), for cloud photographs; the Art Department of *The New York Times,* for newspaper weather maps; Simon and Schuster, Inc., Publishers, for permission to quote excerpts from letters by Pascal and Périer from *A Treasury of the World's Great Letters,* edited by M. Lincoln Schuster, Copyright, 1940, by Simon and Schuster, Inc.; to the Proprietors of *Punch* and the author, for permission to quote an excerpt from a poem by Sir Alan P. Herbert; to Professor Glen H. Mullin, Columbia School of General Studies, for counsel in preparing the manuscript; to Professor Herman F. Otte and Robert B. Hutchinson, for suggestions; and to Gwladys M. Davies, for care and patience in typing the various drafts of this book.

34 Crane Road
Scarsdale, New York
1951

What is it moulds the life of man?
The weather.
What makes some black and others tan?
The weather.
What makes the Zulu live in trees,
And Congo natives dress in leaves
While others go in furs and freeze?
The weather.

—Anonymous

Contents

Illustrations

1

How to Make Your Guess Better than Mine

Hath the rain a father? or who hath begotten the drops of dew? Out of whose womb came the ice? and the hoary frost of heaven, who hath gendered it? . . . Canst thou bind the sweet influences of Pleiades, or loose the bands of Orion? . . . Canst thou lift up thy voice to the clouds, that abundance of waters may cover thee?

—*Job*

A DETACHMENT of weathermen in Italy, so the story goes, lost all of their instruments when a shell demolished their specially built jeep. After a desperate search for new equipment, one of the men came across a donkey that quickly made a reputation for being the best forecaster in the group. Whenever he brayed for an unusual length of time, the soldiers would notify headquarters of impending rain. The donkey's predictions proved extremely satisfactory until one day a female donkey came along. Thereafter, no one could tell whether the prediction was for "Rain" or "Fair and Warmer"!

This story may confirm many suspicions concerning the similarity between the weatherman and a certain four-legged animal. It is intended to show instead that you needn't have a jeepload of instruments at hand to begin to forecast the weather. In fact, even a donkey isn't absolutely necessary. For, by becoming familiar with some of the forces of the atmosphere which make the weather, by interpreting weather data printed in the daily papers, and by watching for the danger signals of the sky, you can foresee significant changes and adjust your plans to them.

The weather always gives advance warning of these changes which you can heed. If quickly moving clouds lower and increase in number toward dinnertime, you may expect poorer weather by breakfast. In summer, billowing white clouds with mushrooming tops, accompanied by static on the radio, are signs of sudden showers. Dark clouds gathering on the horizon to the west warn of rain or snow. A rapidly rising barometer and a steady wind from the northwest indicate clearing skies. A cloudless winter sky at sundown means an extra blanket on your bed before sunrise. But a cloud-filled summer sky in the evening tells of a hot and sleepless night to follow. When the setting sun looks like a ball of fire and the sky is clear, or when the moon shines brightly and the wind speed is light, tomorrow will be fine.

Watch for unsettled weather when clouds move in separate directions at different heights. Look for rain when they huddle around mountain tops and begin to descend down the sides. There are people who also look for rain whenever a cat sits with her tail to the fire. These prophets are not to be regarded lightly, because sooner or later rain does fall and their forecast is verified.

In order to follow the weather, you don't have to buy expensive machines or build complicated gadgets. But if you want your guess to be better than the neighbor's, you should buy a daily newspaper that contains a weather map as well as other data. For sixty cents a month, you can get the daily weather map prepared by the Weather Bureau which will be helpful (see Appendix 3).

In the line of instruments, a thermometer is handy. It is worth having even though you glance at it only once a day when deciding how to dress in the morning. You can buy a thermometer that is reasonably accurate at the nearest drugstore for a dollar or less. When you install it, put the instrument outside a window facing north so that as little sunlight as possible will fall on it.

Or for nothing at all, catch a cricket and use it as a chirping thermometer. Crickets, it has been found, chirp at a rate which increases as the air temperature rises—the higher the temperature, the greater the number of chirps per minute. Provided the thermometer reads between 45 and 80 degrees, you can count the number of chirps in 15 seconds, add 37, and come out with the correct air temperature!

To measure the pressure of the air, a simple box-type

barometer is desirable but not indispensable. You can purchase one at any price from $5 to $50. Keep the barometer in a dry corner of the house which stays at about the same temperature all day. However, you can buy several contraptions at the dime store that will serve as a fair substitute for a barometer, such as a narrow glass tube filled with colored water and shaped like a swan's neck.

For wind speed and direction, all you need is a chimney, flagpole, or tree. Using the table in Appendix 1, you can learn to estimate wind speed by the eye with little difficulty. And you'll soon discover, as Francis Bacon did, that "Every wind has its weather."

A saltcellar will serve as a rough gauge of the humidity of the air. If the salt comes out easily, the humidity is low. If it refuses to pour, the humidity is above average. And when your clothes cling to your skin and perspiration runs down your neck, the humidity is beyond description.

Finally, as an aid to identifying clouds, the Weather Bureau puts out an excellent illustrated pamphlet which you may buy for thirty cents (see Appendix 3).

With these tools in hand, you can start to outguess your neighbor. Is the wind increasing in speed from the south? Are the clouds becoming thicker and lower? Set out your umbrella and take it to work with you in the morning. Or does your newspaper weather map show clear weather to the west which may move overhead in the next twenty-four hours? If so, plan to take the children on a picnic in the afternoon—but bring your raincoat along just in case.

Before you begin to forecast the weather more seriously, however, you must first investigate some of its whys and wherefores. For there is more to weather than a sharp eye, a sensitive toe, a cricket, and a saltcellar can tell. As the next chapters disclose, the course of your search will reveal how almanacs are compiled. It will show why the California Chamber of Commerce is indebted to the Pacific Ocean. You'll discover that the atmosphere is built like an onion, and that the Atlantic High is not a cocktail. You'll study the telltale clouds and learn that some are made of water and others of ice. You'll become acquainted with the chinook, a wind that eats snow, and the tornado, a wind that kills men. You'll find that the sky is a battleground in which perpetual war is waged between huge armies of air. Above all, you'll gain an understanding of the habits of the weather and thus be able to anticipate its major changes yourself.

To get the most out of weather, then, you need to combine sensitive toes with sensible theory. With the aid of both, you can follow the weather with confidence. You won't, of course, ever force the Weather Bureau out of business or "loose the bands of Orion." But you certainly can make your guess better than mine and disprove the English proverb that talk about the weather is a "discourse of fools."

2

From Noah to Now

We may expect some showers of rain this month or the next, or the next after that, or else we shall have a very dry spring.
—February forecast from *Poor Robin's Almanac*, 1664

FORECASTING the weather is one of the oldest of all occupations. The first man in this business was Noah, who built an ark of gopher wood because he confidently anticipated forty consecutive days of rain. And, unlike some more recent weather prophets, Noah lived to see his prediction come true.

Joseph of Egypt was even more successful than Noah. As an interpreter of dreams, Joseph anticipated Freud. As a meteorologist, he has never been surpassed. No one today, for instance, could say with confidence, as Joseph did, "Behold there come seven years of great plenty throughout all the land of Egypt: And there shall arise after them seven years of famine. . . ." To predict rain and abundance, or drought and famine, fourteen years in advance is not even attempted by the Weather Bureau. It does forecast rain seven days ahead of time. For longer periods, however, its weathermen know better than to try to translate dreams in terms of precipitation and nightmares in terms of cold waves.

Ever since Joseph made his fortune by foretelling the weather, there have been people who have tried to imitate him. Countless dealers in black magic and self-educated rain doctors have paraded through history. Each one has used his own special method of deceiving people into believing that he could do something about the weather.

In the Middle Ages, for example, the lords of the land had astrologers attached to their courts whose job it was to read the stars and thereby tell the weather. These astrologers prepared almanacs for the coming year. Their almanacs gave forecasts of the weather, as well as of births and deaths and fortune or misfortune on the battlefield. The astrologer naturally would prophesy conditions most agreeable to his master. If the lord liked to spend December in hunting, the stars would indicate favorable weather for that period. An astrologer in an adjacent castle, nonetheless, might specify snow and cold weather for December if his earl preferred to sit by the fire and feast the month away.

The almanac, notwithstanding, was consulted religiously by everyone. If the mariner wanted to find out what the weather would be for his coming voyage, he turned either to an astrologer or to an almanac. If he were learned, he might also read Aristotle's *Meteorologica*. This book was the unquestioned authority on weather theory for a period of 2,000 years after it was written in about 350 B.C. Unfortunately, most of the theory was not only conjectural, but also incorrect. For example, Aristotle explained that "hurricanes generally are formed when some winds are blowing and others fall on them."

Since this nonsense was of little help to the seafarer, he might try *The Book of Signs,* compiled in about 300 B.C. by Theophrastus, Aristotle's friend and successor. This work contained rules for forecasting the weather from a number of signs. Among the forecasting rules were 80 for rain, 50 for storms, and 45 for winds, most of which were undependable. Theophrastus, however, did base a few of his rules on reliable scientific observations. He noted that "streaks of clouds moving from the south mean rain within three days," and "when there is fog, there is [often] little rain."

Many books of a similar nature appeared after that time. None of them was of much practical value. But no matter how false their rules and predictions, they were better than nothing at all.

In the fifteenth century, one book encouraged the reader to forecast the weather for the following year by slicing six onions in half, each onion half thus representing a certain month of the year. Then, before the reader left for midnight mass, he was supposed to salt the onions liberally. Upon returning in the morning, he could judge the wetness of each month in the future by noting the condition of the salt on each half. The farmer, on the other hand, could expect rain when his pigs came home with straw sticking out of their mouths. The theory in this case was that the pigs knew that rain would soon fall, and they rushed to their pens in such a hurry that they hadn't finished eating.

So it was little wonder that one of the best-sellers of the year 1555 was *The Prognostication of Leonard Digges*. In his book, Digges affirmed that "The Moone, in Cancer, Leo, Capricornus, or Aquarius, ayded with any aspect, but chiefly with the opposition or quadrate of Venus, rayne folowyth." Anyone with an elementary knowledge of astrology could realize the truth of that statement. And the author, of course, was sly enough not to state just how soon "rayne folowyth"!

Thus, from Noah to the latter half of the nineteenth century, people had little more than guesswork and folklore to give them an indication of what the future weather would be.

Scientific weather forecasting was started shortly after the year 1844, when Samuel Morse sent the world's first telegraphic message over a line from Baltimore to Washington. Up to this time, it had been impossible to collect weather data quickly. Observations were transmitted by letter, and since mail moved more slowly than weather, data were usually obsolete before they were received. The telegraph revolutionized this situation by enabling the weatherman to assemble and interpret observations within an hour after they were taken throughout the country.

Aided by the telegraph, the modern weatherman has made rapid progress in forecasting, especially in the last forty years. Today, instead of relying on the "quadrate of Venus," he prepares an endless number of charts. These charts picture temperature, pressure, and moisture conditions at many levels of the atmosphere. From his charts, the weatherman reads "clear with little change in temperature" more easily than the hoary astrologers read their horoscopes.

As a result, when General Eisenhower wanted to know what the trend of the weather would be during the invasion of Europe, he didn't offer up a sacrifice to the gods or refer to the signs of the zodiac. He depended instead upon the skill of men trained to gather data and forecast scientifically. He did not misplace his trust. Despite their forecast of unfavorable weather, General Eisenhower decided to launch the invasion anyway to take advantage of other more favorable conditions. Had he been alive two hundred years earlier, the heavens might have been his only recourse.

By 1850, comprehensive records of weather observation taken with instruments were being compiled. These records aided weathermen in developing new forecasting techniques. They also served to indicate what the climates of the different parts of the world were.

To the ancient Greeks, the word "climate" meant "slope of the land," for they found that there was a change in the wind direction and temperature and moisture condition of the air as the earth "sloped" away from the equator. Consequently, the climate of Greece denoted to them (as it does to us today) the *average weather* conditions that prevailed over that country through the years. Moreover, we now have records of observational data to demonstrate these facts, whereas the Greeks could rely only on hearsay.

It is the climatologist who collects these records of monthly, seasonal, and yearly variations in pressure, temperature, cloudiness, precipitation, wind, and other weather elements. From his data for the United States, he can characterize the various climates of the country, as

S. D. Flora has done, for example, in the *Climate of Kansas:*

> Kansas, the premier wheat state, has an annual mean temperature almost as high as that of Virginia, more sunshine than that of any state to the east, and generous summer rains which, in the eastern counties, average heavier than those of other states, except a few along the Gulf Coast.
>
> This favorable combination of weather elements and availability of more arable land than that in any other state, except Texas, accounts for the high rank of Kansas in crop production, finished livestock, and dairy products.
>
> The State lies across the path of alternate masses of warm moist air moving north from the Gulf of Mexico and currents of cold, comparatively dry, air moving from the polar regions. Consequently, its weather is subject to frequent and often sharp changes, usually of short duration.
>
> Summers are inclined to be warm—often the word "hot" describes them best—but are healthful, with low relative humidity during periods of high temperatures, and usually a good wind movement. Heat prostrations are almost unknown. Summer nights are usually cool, especially in the western counties.
>
> Winters are drier, with more sunshine than those of eastern states. The average snowfall is less than that of other states, except those located farther south. Michigan, Pennsylvania, New York, and the New England States normally have from two to three times as much snowfall as Kansas.

Or the climatologist may draw up charts like the one prepared by the Weather Bureau depicting the "Average Annual Number of Clear Days" (Plate 1).

If you live in Seattle, Washington, you'll find by studying this chart that the sky is clear only 80 days a year on the average. In Yuma, Arizona, the average is 280 days. So while the Washington State Chamber of Commerce may publicize its salmon fishing, the Chamber of Commerce of Arizona can rightfully proclaim that Arizona is the sunniest state in the country.

The climatologist would interpret this chart in the light of many facts. These would include the distance of the locality in question from the equator, the elevation and slope of the surrounding land, the proximity of sources of moisture like lakes and oceans, the flow of ocean currents, and the prevailing wind direction.

He could show that "the sunny southwest" has the

greatest number of clear days because the air is drier in that part of the country than anywhere else. This is the result of the fact that the surrounding mountains prevent moisture-laden air from reaching the area in most months of the year. In addition, the prevailing circulation of air strongly favors a dry climate. Moreover, the sun warms the air to such an extent that a larger amount of moisture than normal is needed to produce cloudiness.

The southern portion of Louisiana, on the other hand, has only 110 to 140 clear days a year. Louisiana is cloudy because the prevailing flow of air over the state comes off the Gulf of Mexico, where the air is moist, and clouds form readily. The number of clear days is greatly reduced in Washington and Oregon. In these states, moist air is often trapped in the valleys during the winter for weeks at a time, so that fog and low clouds persist for long periods. In New England, clear skies are infrequent because much of the bad weather that crosses the country eventually passes through that region.

The climatologist, therefore, exists by virtue of weather records gathered in the past. His fare consists of the averages and variations of data. The greater the number of years they include, the more he relishes them, for he can use these data to make a climatic forecast. If, for example, he examines forty years of data for Memphis, Tennessee, for October 25, he will discover a weather condition that occurred most frequently on that date. He then concludes that this *average* condition (or some other one that proves most probable) should be more likely to prevail on the next 25th of October than any other. Having made this decision, he issues his climatic forecast accordingly.

On the basis of these climatic values, many popular almanacs are prepared. Since average weather conditions seldom change, however, an almanac prepared exclusively from such data would be the same from year to year. So to give the weather more variety and to keep customers coming back for new almanacs, each year some publishers change a few entries.

The *Old Farmer's Almanac* made its reputation in this way by "changing the weather" unintentionally. When one of the typesetters was given the proof of the *Almanac* for the year 1816, he noticed that the entry for the weather on July 13 had been omitted. Upon asking the busy editor what the entry should be, he was told sharply, "Put anything there you please." The typesetter then decided to note something more interesting than "Clear" or "Hot" for July. He entered what he considered to be the most unlikely weather he could conceive for that date: "Rain, Snow, and Hail." And when rain, snow, and hail fell that July 13 in a freak storm, the *Almanac's* reputation was established.

In such a manner, climatological data are used in making almanacs. They are put to work in many other ways as well. Construction firms can determine the most favorable months for building in various regions. Airfields can be located where fog occurs least often. Motion picture companies can select outdoor locations having the most sunshine and the least cloudiness.

The climatologist can also use his data to explain the existence of various types of landscape. For there is a close connection between climate and landscape. In regions where average temperatures are high and precipitation is abundant, for example, erosion of the land by water is active. Laced by a network of streams, the countryside tends to be gently rolling and well covered by vegetation. Forests are dense. Soils enriched with decayed matter from animals and vegetation are usually dark in color, fertile, and easily tilled.

In regions of high temperature but little rainfall, plant growth is discouraged by the lack of precipitation and the strong alkalinity of the soil. Unprotected by a solid blanket of vegetation, the countryside is scoured by the action of the wind. And since the landscape is not subject to the leveling effect of water erosion, it is characterized by sharp breaks and steep slopes, and alternating plains and plateaus. A similar landscape is evident in regions of low temperature and little rainfall, although the soil, being frozen, is scarred more by glacial than by wind erosion.

Yet the collection and interpretation of climatological data are only one branch of meteorology. When the airlines in Denver want to know what weather conditions may be expected in Chicago in the morning, they need more than climatological charts to give them this information. These charts are helpful in indicating what *average* weather condition can be expected, or what extremes could possibly occur. But they tell only what the weather would be if it were behaving normally. If it is not—and much of the time it varies one way or the other from the average—then the climatological data must be supplemented by the charts of the weatherman.

The weatherman uses these charts, which picture the state of the weather over the country at a given moment, plus his knowledge of climatology to make a forecast of "the weather." He should be able to tell the flight dispatcher that the bases of the clouds at Chicago are expected to be 1,500 feet above the ground, slowly improving to 4,000 feet by afternoon. This condition may be exactly what climatological records would indicate. Or it may be exactly the opposite. Whatever the case, the weatherman is concerned most with the hourly and daily variations in the climate which constitute the weather.

In winter, for example, he must keep a close watch on the temperature. If he forecasts it to fall below freezing so that streets turn from slush to ice, hospitals will call out extra ambulances. Highway departments will use more men and sanders, and garages will alert their wreckers. A forecast of daily temperatures below zero means that the railroads must prepare for trouble. As George H. Baker, general superintendent of passenger transportation for the New York Central Railroad, reported to a *New York Times* interviewer in February, 1948:

> When the mercury hovers near zero . . . more heat is necessary to keep passenger cars and diners comfortable. This draws more steam from the locomotive, and when the steam pressure falls under the increasing demand, the train has to sacrifice its speed for passenger comfort. Then schedules begin to jam. Frequently in such weather, trains are held down to sixty miles an hour to keep the passengers comfortable, although this may confound the timetable.
>
> Coal freezes in the engine tender and the automatic stokers are apt to balk or break down. Then an unscheduled stop may be necessary to take aboard an extra fireman's helper to break up the coal or even to help hand-fire the boiler.
>
> Diesel-electric locomotives also suffer from the cold. It takes more steam to keep the passengers warm; this uses up water faster and extra, unscheduled stops are required to refill the boilers. One cold

morning recently the Twentieth Century Limited with 9,000 horsepower in the Diesel-electric on its headend arrived two hours late.

To forecast these temperature changes, and related changes in pressure and moisture, the weatherman collects hourly weather reports from every part of the country. He interprets them in the light of past experience and present trends. He bases his daily forecast on the projected change in temperature, moisture, and circulation of air. For it is the change in these three elements which is responsible for all the weather that has occurred from the time of Noah to now.

Forecasting the weather begins, therefore, by discovering how the temperature, moisture, and air circulation change and what processes cause them to do so. To become acquainted with these elements, we must be introduced first to the atmosphere, since it is the medium in which all weather is created. Once we have seen how our "airy shell" is constructed, it will be easier to understand how the weather moves within it.

3

The Atmosphere: the Largest Unexplored Region in the World

> This most excellent canopy, the air, look you, this brave o'erhanging firmament, this majestical roof fretted with golden fire. . . .
>
> —Shakespeare, *Hamlet*

THE LARGEST unexplored region in the world is not located at the poles, in Alaska or Tibet, nor beneath the oceans. It lies above us, in the atmosphere. Seventy-five times as deep as the deepest ocean, and ninety times as high as the highest mountain, the air overhead extends roughly 500 miles aloft (estimates range from 300 to 700 miles). Ninety-seven per cent of this space is unexplored by man himself. The remaining 3 per cent was penetrated by Captains Anderson and Stevens in 1935, when their balloon, the *Explorer II,* rose 13⁷⁄₁₀ miles above sea level (72,395 feet) over the South Dakota hills.

It is an overnight trip on the train between New York and Detroit. The distance of 500 miles between these two cities may be traveled with ease. No one would give a second thought about what to put in the overnight bag, or how strongly the Pullman was constructed.

But suppose that Detroit were 500 miles upward. Then trouble begins. If your Pullman weren't well heated, even the warmest clothes would be useless, for you might pass through temperatures of 100 degrees below zero. If it weren't well air-conditioned, you'd be overcome later by temperatures close to boiling. Moreover, if the car weren't well sealed and pressurized, you would lose consciousness before rising 25,000 feet.

To fly in the upper atmosphere, you'd have to exchange the Pullman for a rocket. Even a conventional type of airplane cannot reach altitudes much greater than 50,000 feet. At those levels, the air isn't dense enough to give the required support or lift. With all these complications, no wonder Captains Anderson and Stevens figuratively stopped at Yonkers, just outside the city limits of New York, instead of going all the way to Detroit!

Unfortunately, scientists have not yet determined the exact height of the atmosphere, as their estimates of from 300 miles to 700 miles indicate. They have, however, traced the flight of meteors at altitudes as low as about 45 miles, where there is enough air to heat the "shooting stars" to incandescence. A rocket plane flying only 45 miles above the earth, consequently, would be in danger of colliding with one of these meteors. The damage from a crash might be considerable.

Scientists have also watched rare noctilucent clouds, which are composed of meteoric or volcanic dust (or possibly ice crystals), and are brilliantly illuminated long after sunset by the sun on the other side of the earth. These clouds form at elevations of approximately 55 miles (Plate 12). The "golden fire" of noctilucent clouds was observed for several years after the violent eruption in Sundra Strait between Java and Sumatra of the volcanic island of Krakatoa in 1883.

Farther aloft, a series of atmospheric layers, extending from about 60 to 200 miles, have been located which reflect radio waves back to earth. In addition, the Aurora Borealis, or Northern Lights, perhaps caused by showers of particles from the sun which strike atoms and molecules of air and make them glow, appears at the same altitudes. It has been seen at an elevation calculated to be about 700 miles.

Special unmanned rockets are now being designed to carry recording instruments to several hundred miles. Perhaps they will eventually reach even the top of the atmosphere. When weathermen can gather enough rocket data from the upper levels of the air and can learn how to interpret them, they will discover much more about the extent of the unexplored atmosphere. They will also probably get some help in developing better methods of forecasting "fair and warmer" for the Chicago White Sox-Detroit Tigers baseball game thirty days from now.

With the help of rockets, balloons, airplanes, radios, telescopes, and educated guesswork, scientists have found that the 500-mile deep atmosphere is constructed like a huge onion. This atmospheric onion is composed of four concentric layers.

Each of these four atmospheric layers is different from the others. Yet none of them is specifically defined or uniform in extent, and the thickness of each layer changes hourly, daily, and monthly. From bottom to top, the layers are called *troposphere, tropopause, stratosphere,* and *ionosphere*. In their balloon flight of 1935, Captains Anderson and Stevens rose through the troposphere and tropopause and investigated part of the stratosphere.

The layer of the atmosphere in which we live is the troposphere. Within this layer, air moves vertically (tropo means "overturning") as well as horizontally. This vertical circulation of air allows moisture from oceans and lakes to be carried to all levels of the troposphere. Whenever the moisture condenses, clouds are formed. They appear in the lowest 20,000 feet as stratified sheets (stratus) or as heaped domes (cumulus).

Above 20,000 feet, clouds are of the cirrus variety. The delicate, white cirrus, which often resemble feather-like plumes drawn across the sky (Plate 2), trace out the upper limits of the first layer of the atmosphere, the troposphere.

Over the United States, the troposphere is about 40,000 feet (nearly eight miles) deep. Since the troposphere is the only atmospheric layer in which there is large-scale vertical movement of air and cloudiness, all the weather of the world occurs within it. Thus in less than 2 per cent of the thickness of the atmosphere, 100 per cent of its weather is manufactured.

Much of the cloudiness which is confined to the troposphere never rises above 20,000 feet. Commercial airliners equipped to fly at altitudes of 20,000 to 25,000 feet can thus pass safely over most areas of bad weather. Passengers don't suffer from airsickness and pilots needn't worry about running into mountains masked by clouds.

If you should ever fly upward about 40,000 feet to the top of the troposphere, take along fur-lined garments. They should be warm enough to withstand temperatures of at least 70 degrees below zero. For, on the warmest summer day, the troposphere will freeze you before you ascend 20,000 feet. It will pack you in dry ice before the end of your trip.

When the troposphere is in its normal state, the average decrease of temperature with height is about 3½ degrees per 1,000 feet. Thus, if the thermometer reads 70 degrees at the airport runway, it will be close to zero 20,000 feet overhead. This value of about 3½ degrees, however, is an average one. It often does not apply in specific cases. Nonetheless, a hailstorm on a hot July afternoon is good evidence that the skies are well air-conditioned.

This fall of temperature in the troposphere may be partially explained by the fact that the earth acts like a stove, giving off to the atmosphere the heat it receives from the sun. The air nearest the earth becomes warmest, while that farther aloft in the troposphere becomes colder and colder. Close to this basement stove—perhaps in your neighbor's back yard—roses bloom throughout the year. But in the attic—three miles overhead—the mountains shiver in snow.

Although the temperature decreases with height in the troposphere, the average wind speed constantly increases. At about 40,000 feet, it reaches a maximum. Consequently, the wind speed is usually greater on top of Mount Whitney in California than at Fresno in the valley below. The ventilation of a room on the fiftieth floor of a New York hotel is better than that on the fifteenth. Writing about the ascent of the *Explorer II* in 1935, Captain Stevens noted that even in the lowest level of the troposphere, the increase in wind speed with height was sufficient to make the inflation of the *Explorer II's* balloon extremely difficult. For although the gondola on the ground might be in calm air, the top of the balloon, 315 feet overhead, was frequently swayed by winds of six to eight miles per hour.

There are days, however, when the winds aloft are weak and far below their average speed. If the sky is clear, such days are fine for "sky writing." For then messages written in the air with smoky ink aren't quickly blown away, as they would be under average upper-wind conditions.

Wrapped around the troposphere is the second layer of the onion-like atmosphere called the tropopause. It is a shallow region of transition which marks the upper limit or "pause" of vertical air motion. In this region, the air temperature remains constant and no longer decreases with height, as in the troposphere directly below. Moreover, the greatest average wind speeds are observed at the tropopause. B-29 pilots, for example, flying over Japan during World War II at altitudes close to the tropopause sometimes reported exceptionally high winds of 300 miles per hour. Above the United States, tropopause winds of 150 to 200 miles per hour are not uncommon.

Directly over the tropopause, about eight miles above sea level, lies the base of the stratosphere, the third layer of the atmosphere. Here the wind is stratified and moves in a horizontal plane (with few exceptions), while the wind speed slows down with height. At the bottom of the stratosphere, the air temperature is close to 70 degrees below zero. But in the first 25 miles of the stratosphere, it increases to at least 30 degrees above zero, and may rise further. In about the next 17 miles, however, the temperature falls to at least 30 (and possibly 100) degrees below zero.

When the *Explorer II* reached its record altitude of 13$\frac{7}{10}$ miles, Captain Stevens discovered that the lowest temperature of the flight had occurred at the tropopause at 40,000 feet. He also noted that as the balloon climbed into the stratosphere, the temperature had risen instead of fallen, as it did in the troposphere below.

On top of the stratosphere, starting at about 50 to 60 miles, is the fourth layer of the atmosphere, the ionosphere, our "radio mirror." Radio waves which strike various levels of the ionosphere are reflected back to earth. As a result, they reach receivers far beyond the horizon of the transmitter. Without such a reflecting surface for radio waves, transmission of broadcasts would be restricted to a "line of sight" or horizon distance, as television is today.

Temperature data collected by rockets in the ionosphere are limited. Yet they reveal that it may be one of the hottest and coldest places in the world. About 50 to 60 miles above the United States, the temperature is at least 30 (and possibly 100) degrees below zero. Then it begins to rise again. At about 70 miles, it reaches freezing, and may even climb far above the boiling point of water at higher elevations. Above this level, there are no reliable data available, for the temperature of the upper ionosphere has not been taken even by instruments (Figure 1).

Throughout the lower layers of the atmosphere, the gaseous composition of the air remains constant. With the exception of carbon dioxide and moisture, there is a uniformity in the relative proportions of gases up to about 45 miles. Below this height, the air is composed by volume of 78 per cent nitrogen, 21 per cent oxygen, and 1 per cent rare gases.

But there is not the same amount of oxygen at each level of the lower atmosphere. The density of oxygen, like the density of air, drops off aloft rapidly. If you've ever gone up Pike's Peak in Colorado (2$\frac{7}{10}$ miles or 14,110 feet high), you may have discovered this fact. For by the time you reached the top, you were probably short of breath, a little dizzy, and had a slight headache—all for want of enough oxygen. If you could have ascended an-

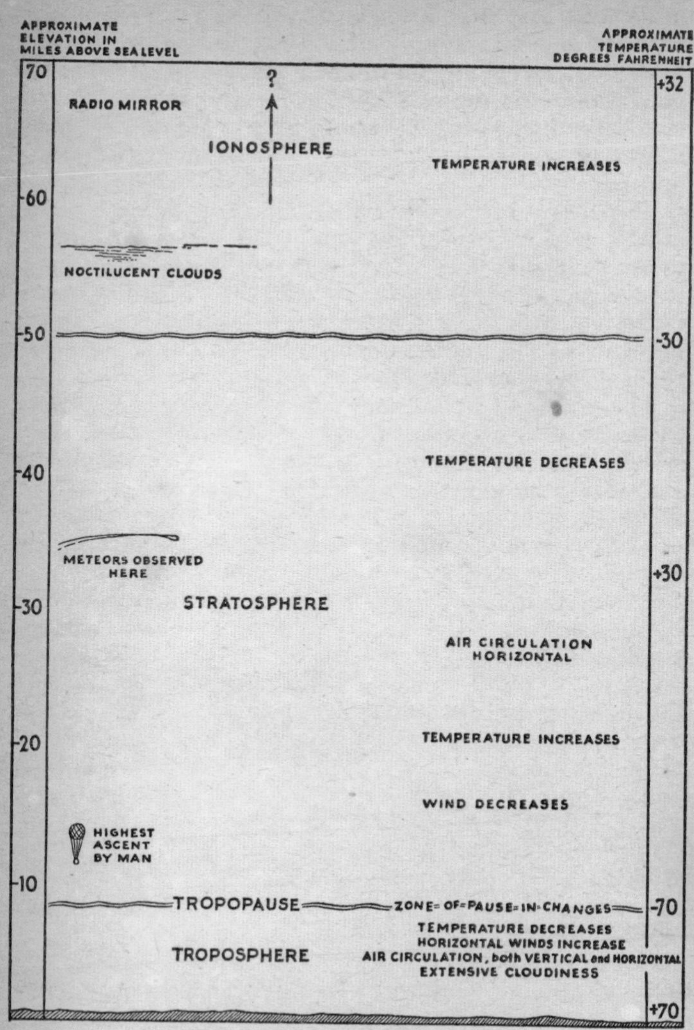

THE LOWER ATMOSPHERE

FIGURE 1

other mile into still thinner air, you would have fainted without an extra supply of oxygen to supplement the oxygen deficiency of the rarefied air. To protect passengers against the harmful effects of such oxygen deficiencies, stratoliners that fly above 10,000 feet have sealed cabins in which the density of the air is kept close to normal.

The oxygen content of the air thus remains *proportionately* constant throughout the lower atmosphere. The moisture in the air, however, varies with height, latitude, season, and time of day. It is nearly all confined to the troposphere, the lowest layer of our atmospheric onion. For only in the troposphere do huge masses of air move vertically and circulate moisture throughout its depths.

The degree of wetness or dryness of air at any level of the troposphere is measured by a percentage ratio known as the "relative humidity." This ratio compares the amount of uncondensed moisture actually present in the air at a given spot and at a given temperature to the maximum possible amount that could be contained by that air at the same temperature.

When the air at a given spot holds all the moisture it can at a certain temperature, its relative humidity is 100 per cent. The air is then "saturated." If its relative humidity is 50 per cent, the air actually contains only one-half the moisture it could hold at that temperature at the same given spot. On an arid day, when the relative humidity of the air is low, clothes dry rapidly on the line and hair crackles with electricity as it is combed. On a humid day, when the relative humidity is high, windows stick in their frames and salt fails to pour from saltcellars.

As the relative humidity increases from next to nothing to 100 per cent, the human hair expands 2½ per cent in length. This fact should be of value to barbers and wigmakers. It opens a vast field of research for some industrious manufacturer of hair restorers. As A. P. Herbert says, it means also that permanent waves go all awry:

> . . . But I know ladies by the score
> Whose hair, like seaweed, scents the storm,
> Long, long before it starts to pour
> Their locks assume a baneful form.

The warmer the air becomes, the more moisture it can contain before becoming saturated. At ground level, an increase of about 20 degrees in temperature doubles the moisture capacity of a cubic foot of air. Saturated air at 60 degrees, for example, contains twice as much moisture as saturated air at 40 degrees. Thus you feel stickier on a warm day with the relative humidity at 80 per cent than on a cold day. For at the higher temperature there is more moisture in each cubic foot of air.

As the relative humidity of the air approaches 100 per cent and the air becomes saturated, watch for the formation of clouds, fog, dew, or frost.

Clouds often develop when unsaturated air is lifted over mountains or colder air, or when the air descends of its own accord after being heated. As the air rises, it cools at the rate of about 5½ degrees per 1,000 feet. (This rate of cooling should not be confused with the over-all average decrease of temperature of about 3½ degrees per 1,000 feet throughout the troposphere.) If no moisture is removed from the air as it cools, its relative humidity increases until either the air stops rising or it becomes saturated and cloudy.

Air that flows downhill, however, *warms* at the rate of about 5½ degrees per 1,000 feet. Consequently, there is less cloudiness, rainfall, and vegetation on the lee side of mountains, where air descends, than on their windward slopes, where air rises. On the windward side, clouds are numerous and rainfall abundant, for the air is lifted, cooled, and becomes saturated as it rises up the slopes. But once it descends to leeward and is warmed, its relative humidity is reduced and the amount of possible cloudiness and rainfall is minimized.

The location of great belts of forest reflects this fact that rainfall is usually heavier on windward mountain slopes. As noted in *Climate and Man:*

> There are two of these belts in the United States, along the Atlantic and the Pacific. In between, farthest from the oceans, and therefore farthest from the primary sources of moisture, the land grows grass, not trees. In the western belt, three successive mountain ranges interfere with the passage of the sea-born moisture eastward. *The west* [windward] *side of each range is moist, and forests grow*

there but not on the drier eastern [leeward] *side.* . . . In the eastern part of the United States, on the other hand, there are no great mountain barriers to stop moisture-laden air, and the forest is continuous.

When the relative humidity of only a few inches of air near the earth reaches 100 per cent, the moisture in the air condenses on the ground either as dew or frost.

In summer, the water is chilled out of the air to "fall" as dew upon the cool earth, just as it forms a dewy "sweat" upon ice-cold water pitchers. The dew "falls" first upon bodies which cool most rapidly. Grass and small plants, for example, are covered early in the evening when conditions are favorable. Those which lose heat slowly, like rocky ground, may not be covered until morning. The evening "fall" of dew is greatest when the sky is clear and when storm clouds are miles away, so that:

When the dew is on the grass
Rain will never come to pass.

In winter, when the temperature is below freezing, frost appears in place of dew. It condenses from moisture in the air to icy crystals which line the windowpanes and whiten the ground. Frost is not frozen dew, as is commonly believed, since the moisture in the air goes directly into ice without turning into watery dew.

Frost, dew, and clouds consequently form and dissipate as the relative humidity of the air changes. This change is tied up with the temperature and circulation of air through 500 miles of troposphere, tropopause, stratosphere, and ionosphere—the largest unexplored region in the world. Within that vast ocean of air, upon which we depend for life as utterly as fish do upon the water in the sea, is conceived the weather of the world.

4

Temperature: Mud Pies and Climate

While the earth remaineth, seedtime and harvest, and cold and heat, and summer and winter, and day and night shall not cease.
—*Genesis*

NO ONE has ever made a success of selling homemade mud pies. For one thing, they're unpalatable and indigestible. For another, mud pies are difficult to cook. You can put one under a broiler on a gas stove, but it won't cook thoroughly. The upper crust may be burned to a crisp; however, because heat is not easily transmitted through the pie, the filling will remain raw and muddy. Mud pies, then, are definitely impractical.

But if you set a dishpan of water in the broiler in place of the pie, the results are different. Water cooks easily; for as the surface is warmed by the gas flame, the heat is distributed slowly through its entire volume. Consequently, the water at the bottom of the pan becomes about as warm as that on the surface. Yet there is a drawback in this case, too. Although water can be cooked, it cannot be sold for pie.

Broiling either mud pies or water, therefore, is a waste of time. Nevertheless, the implications of the experiment are important as far as the weather is concerned.

Suppose, for instance, that the mud pie were the continent of the United States, the dishpan of water the oceans around it and the gas flame the sun. When the sun rises, what happens? The surface of the earth, like the upper crust of the mud pie, warms rapidly. All the heat it receives from the sun is absorbed within the first two feet of topsoil. But the huge dishpan of water, the ocean, warms little. For the sun's rays are distributed through a large volume of water, all of which must be heated before the temperature of the water at the surface changes perceptibly. By midafternoon, consequently, the surface temperature of the land may have risen 10 degrees. That of the ocean may be less than 1 degree higher.

After sunset, the process is reversed. The earth cools quickly, since all its heat is close to the surface and escapes easily. The ocean, on the other hand, remains at about the same temperature. The heat it gives off during the night is an infinitesimal part of what it has stored up in its depths. As a result, the twenty-four-hour surface temperature variation may equal 20 degrees for land but only 1 degree for water.

Because the mud-pie earth cannot be cooked to any great depth—and because the dishpan ocean can—a blizzard in San Francisco is as rare as a winter without snow in St. Louis. The westerly winds which sweep over San Francisco blow off the Pacific Ocean, whose temperature is fairly constant the year round. The average monthly air temperature at San Francisco varies from about 50 degrees in winter to 60 degrees in summer. A heavy snowfall is as unusual as a penguin at the equator. But directly to the east lies St. Louis, far removed from an ocean which evens out the seasons by being cool in summer and warm in winter. The people of Missouri, surrounded by a continent of land over which the air temperature fluctuates from one extreme to the other, alternately freeze and perspire as average monthly air temperatures range from about 30 to 80 degrees.

While Missourians stew in their own climate during the summer, vacationists at Atlantic City are air-conditioned by means of the "land and sea breeze." It is another result of the unequal heating and cooling properties of mud pies and dishpans of water. This breeze pushes sailboats on-shore in the daytime and off-shore at night. Like a gigantic electric fan, it circulates cool air over sweating workers and restless sleepers.

At the seashore, you've probably noticed how quickly the temperature of the sand changes throughout the day. In the morning, the beach may be pleasantly cool. But by afternoon it's often so hot that you hesitate to walk on it with bare feet. After sunset, the sand may be so chilly that you have to build a fire to keep warm. Meanwhile, the temperature of the ocean stays about the same, except for variations due to a strong wind shift or a turn in the tide. As a consequence, the "land and sea breeze" is born.

When the sand becomes hot, the air above it is also heated. As the air gets warmer, it expands and becomes lighter and rises upward. At the same time, cooler air from the temperate ocean flows inland to take the place of the warm air. This cooler "sea breeze" air soon makes the temperature fall.

At Chicago, on hot summer days a "sea breeze" frequently blows off Lake Michigan with welcome effects. One July afternoon in 1901, within sixty minutes after the breeze began, the temperature dropped 18 degrees—from 102 to 84 degrees!

The "sea breeze," which rarely comes more than 50 miles inland, may blow as long as the land remains warmer than the ocean. When the sun goes down, the breeze dies away. Later in the evening a "land breeze" may set in. This off-shore wind will begin to blow after the land loses heat rapidly and becomes cooler than the ocean. The "land breeze" forms when cold land air moves toward warmer air over the ocean.

Away from the coast, where there is no "land and sea breeze," temperatures may be disagreeably high in the daytime. For the absence of a steady breeze means that the air warms quickly. And, generally speaking, the lower the wind speed, the higher the day's maximum temperature.

Air that is well mixed by a steady breeze is not so subject to extreme changes in temperature as air in calm regions. When air is well mixed, it warms slowly; like the water in the ocean, a deep layer of air must be heated before its temperature near the ground changes noticeably. But when the air close to the ground is undisturbed by wind, only a shallow layer of air absorbs the earth's heat, and the temperature variation is large.

Consequently, you can expect the air temperature to be highest on the calmest days. For this reason, you're more likely to get a heavy sunburn on a breezy summer day than on a calm one, even though you bask under the same amount of sunlight both times. On a windy day, you aren't so aware of the danger of overexposure. The breeze keeps the temperature of the air down and cools you off by blowing away the heat of your body. Although you may feel comfortable, however, the sun keeps burning its brand deeper and deeper into your skin. Before you know it, your nose is raw and your back is welldone. When the air is calm, on the other hand, its high temperature, plus that of your body heat which is not blown away, warns you to protect yourself or take the consequences.

The daily temperature range in a state, a city, or in your back yard, therefore, is influenced not only by the proximity of dishpans of water like the Great Lakes or the oceans, but also by local winds such as the "land and sea breeze." And because the yearly climate is to a large extent the average of daily temperature ranges, it also reflects these influences. Thus the "public climate" of New Jersey is milder than that of landlocked Missouri. So, too, the "private climate" of the back yard of an ocean-side home is milder than the "private climate" of the back yard of a home 60 miles inland, beyond the reach of the "land and sea breeze."

These differences in daily temperature range and in yearly climate depend likewise upon exposure to sunlight. The "private climate" of the northern slope of a mountain is more severe than that of the southern slope; in springtime, snow remains on the northern slope long after it disappears from the southern one. On a smaller scale, the severity of the "private climate" of your home also depends upon the slope of the land on which it stands. Even at the noon hour with the sun shining, variations in "private climate" arising from varying land slopes may be extreme.

Suppose, for instance, your home is located in Ohio. If the back yard is *level,* on a clear day at noontime in late winter it will receive an amount of sunshine normal for that date. If, however, your back yard slopes moderately toward the *north* at an angle of 1 foot in 12, it will receive a less-than-normal amount of sunshine. In fact, it will face the sun in about the same way that a level plot does in Ontario, Canada. Consequently, the "private climate" of your sloping back yard will be as wintry as the "private climate" of a level back yard in Ontario.

On the other hand, should your back yard slope *southward* at an angle of 1 foot in 12, it will receive more sunshine than normal. In this case, your back yard "private climate" will be similar to the "private climate" of a level back yard in warm, springlike South Carolina.

The slope of land, therefore, influences the "private climate" of back yards as much as latitude influences the "public climate" of entire states.

Hence your home—and the home of your neighbor—has its own exclusive "private climate." This "private climate" is part of the "public climate" of your city and state, and yet differs in various respects from it. One difference, as discussed above, results from the fact that homes are built upon varying *slopes* of land. Another difference comes from the fact that homes are built upon varying *elevations* of land.

For example, if your home stands at the foot of a hill, it will likely be chilled by cool air at night. Cool air always drains into the lowest-lying ground. A neighbor's home 200 feet up the hill, however, may not be touched. As a consequence, your "private climate" may feature cool summer nights, early frost, and large winter heating bills. His "private climate," 200 feet higher, may mean warmer summer nights, later frost, and fewer gallons of oil each winter for the furnace.

Moreover, you'll probably live in a damper climate than he will, for shallow fog, 50 to 100 feet deep, forms readily in pockets of cold air at night over low-lying ground, while higher, warmer land remains uncovered. Hence your home may be cloaked in fog when his is clear. And after sunrise, your neighbor's home will be warmed by the sun long before yours emerges from its shroud of cold, damp fog. The result may be sinus trouble for your wife and frequent coats of paint for your home's heavily weathered exterior.

These variations in "private climate" fluctuate with changes in the weather. When the sky is clear and the wind speed light, they are most noticeable. In poor weather, they are least obvious. For instance, the magnitude of local temperature variations caused when sunshine falls on varying slopes and elevations of land is considerably reduced when the sky is filled with clouds. With an overcast sky, only 20 per cent as much sunlight reaches the ground to warm it up as on a cloudless day. An overcast day, therefore, is cooler and darker than a clear one. An overcast night, however, is warmer, because clouds blanket-in the heat the earth gives off and keep the temperature up.

Consequently, if you forecast a clear sky for the next twenty-four hours, look for warm days, cold nights, and wide variations in "private climate." If you expect the sky to be overcast, look for cold days, warm nights, and small local temperature variations.

Your forecast for cloudiness may also have some effect on the time of occurrence of the day's highest temperature. Generally, it comes about 3:30 P.M., though the time of maximum heating by the sun is at noon, when the sun is highest in the sky. The "lag" between the time of maximum heating by the sun and the maximum temperature of the air is the same that takes place when you kindle a fire in your living room fireplace. While the heat thrown off by the fire rapidly reaches its maximum intensity, the living room warms slowly. The "lag" may be such that the room will be warmest when the fire is dying away.

The earth, like your living room, also takes time to warm up under the sun. It thaws out so slowly in spring that the warmest summer month is not June (when the sun is farthest north), but July. In some parts of the country, there's a "lag" of two months. Moreover, autumn is a warmer season than spring.

Although there's little "lag" connected with the average time of daily minimum temperature (which comes usually at sunrise, after the ground has been cooling throughout the night), the earth is subject to a winter temperature "lag" similar to the one in summer. Thus the coldest month of the year is January instead of December, and February is often the second coldest month.

For ideal living conditions, research has indicated that the day's maximum and minimum temperatures should center around 65 degrees. Any variation upward or downward from this figure means trouble. In most cases, low temperatures don't have such harmful consequences as high temperatures. Surveys have shown, for example, how low and high temperatures influence the deportment of school children. On cold days, their behavior is, relatively speaking, excellent. On warm, humid days, it's impossible!

If you watch your own behavior, you'll find that you, too, probably react to temperatures in the same way. On cool, brisk November days, you're full of energy and optimism, and make few mistakes in your work. But in July, when the thermometer stands at 95 degrees, your work is as unreliable as your temper.

Contrary to popular belief, the temperature of the *air* (which lowers and raises your temper) is the same whether taken in the shade or in the sun. The air temperature *cannot* be 85 degrees in the shade and 110 degrees in the sun, even though a thermometer placed in the sun records a higher temperature than one in the shade. An unshaded thermometer, however, indicates *not* the air temperature, but the higher temperature of the glass and mercury, which become hotter under the sun's rays than the air does. For a similar reason, you feel warmer in the sun than in the shade. This is so, not because the air itself is warmer, but because your clothing and skin, like the glass and mercury of the thermometer, are warmed by the sun to a temperature above that of the air.

As far as air temperature records go, September 13, 1922, was the hottest day on earth. A thermometer in Azizia, Libya, North Africa, stood at 136 degrees in the shade on that day. In the United States, the highest temperature ever observed—134 degrees—was recorded at Greenland Ranch, Death Valley, California, on July 10, 1913.

The world's record minimum temperature of 90 degrees below zero occurred in Verkhoyansk, Siberia, on February 5 and 7, 1892. (Recently, the Russians announced an unofficial low record of 108 degrees below zero, registered at Oimekon, Siberia.) For the United States, the lowest reading, 66 degrees below zero, was made at Riverside Ranger Station, in Yellowstone Park, on February 9, 1933.

These records, along with millions of less spectacular ones, indicate (as the mud-pie experiment suggests) that the distribution of temperature over the United States is uneven. Some parts of the country are close to the dishpan ocean; others are many miles distant. Some parts are shrouded in fog or covered with snow much of the year; others are free of clouds and situated close to the Tropic of Cancer. Even those cities which are at the same latitude do not fare alike. The sun may shine as intensely on San Francisco as on St. Louis. But the responses it produces are not the same. If the July sky above San Francisco is clear and the wind strong from the west, the Californians will no doubt complain about how chilly it is, and wish that they were in St. Louis where the summers are warm, at least. But under that same sun, the people far to the east are mopping their brows, succumbing to heat prostration, and waiting for August to come so they can take a vacation in California.

5

Clouds: Mares' Tails, Ivory Towers, and Tattle-Tale Gray Sheets

Wet weather seldom hurts the most unwise;
So plain the signs, such prophets are the skies.
—Vergil, *Georgics*

APOLLO, the god of the Greeks and Romans, was the patron deity of the famous oracle of Delphi in Greece. The soothsayers of Delphi revered Apollo as a healer of the sick, a poet and musician, a lawmaker, and a warrior, as well as a founder of cities, a protector of agriculture and vegetation, and an ideal of manly beauty. But above all, they worshiped him for his insight into the future. As the mouthpiece of Zeus himself, Apollo revealed coming events through various signs interpreted by the priests of the oracle. Among the many signs they studied were those disclosed by the clouds, believed to be Apollo's own cattle sent out to graze "in the meadows of heaven."

Although they may not have known it at the time, these soothsayers couldn't have chosen a better omen than the clouds to announce the future of the weather. For the clouds reflect every change aloft in pressure, temperature, and humidity. Not only do they speak of rain and snow, or thunder and lightning, but also of fair days ahead. With their various sizes, shapes, colors, directions of movement, elevations, rate of growth or dissipation, amounts, speed, and time of appearance, the clouds are the bulletin boards of the atmosphere.

To make one of these clouds is not difficult. All you have to do is to close the door of your bathroom and turn on the hot water in the shower. If the air in the bathroom is chilly, the warm, moist air in the shower will mix with it, cool to its saturation point, and condense its moisture into a fog-like cloud. At the same time, more of its moisture will condense on the colder mirror and washbowl and dampen the walls.

The atmosphere likewise generates clouds by the bathroom method of cooling moist air until it is saturated and then condenses. As in the bathroom, a body of moist air can be cooled by mixing it with another body of chillier air. It can also be cooled by pushing it northward over colder ground. It can be cooled in a third way by forcing it to rise over mountains or over a mass of heavier air. Every 1,000 feet the air rises in this manner, it loses about 5½ degrees of heat.

This recipe for making clouds still lacks an essential ingredient. Before moisture in saturated air can condense, it must have something to condense on. Bathroom mirrors are excellent, but they're too expensive an item for the whole atmosphere. Instead, the moisture prefers small particles of sulphur and nitrogen compounds (present chiefly in smoke). It is also partial to sea-salt particles, which escape from the ocean surface and drift far inland. The quantity of such particles in the air depends upon the wind velocity, the proximity of the ocean and of large industrial areas which give off huge quantities of smoke, the temperature and moisture properties of the air, and the time of day. These particles are so small that they're imperceptible under an ordinary microscope. Yet they are essential to the formation of clouds.

The mortality tables for clouds are full of figures, for the life span of a cloud is highly complicated. Just as the cloud in your bathroom will disappear once the water is turned off, so the clouds in the sky will vanish if they aren't fed regularly with moist air currents. These air currents are best developed during the day, when the sun heats the land and water to different degrees of warmth. Hence some clouds tend to grow up quickly in the daytime and perish from thirst at night. Other types of clouds prefer to sprawl out in the evening when the air close to the ground is cooled to saturation, and then melt away after sunrise.

Nearly all the low-flying clouds you see below an average elevation of 20,000 feet are of the bathroom type, made of water droplets suspended in the air. These water clouds range from *cumulus* puffs of cotton to blankets of *stratocumulus* to tattle-tale gray sheets of *stratus* or fog to veiled *altostratus* to corona-ringed *altocumulus.*

Cumulus clouds (Plates 2, 3, and 4) are dense clouds with slight vertical development. They never overcast the sky. Their bases, which form at an average elevation of about 1,600 feet (considerably higher in Plate 3), are almost horizontal. Their upper surfaces are dome-shaped. (*Cumulus* comes from the Latin word meaning "heap.") Cumulus clouds occur most frequently on warm summer days. They often form in the morning and disappear in the evening. Generally, if cumulus clouds appear alone in the sky, they are precursors of fair weather.

In the late afternoon of a hot summer day, as cumulus clouds begin to dissipate, a layer of stratocumulus clouds may develop from them (Plate 4). In other instances, stratocumulus may be found in the vicinity of thunderstorms, and both in advance of and behind extensive storms. Stratocumulus clouds are a continuous sheet or patches of clouds composed of rounded masses or rolls, which are soft and gray, with darker parts. The rolls are often so close together that their edges join. The bases of the clouds are always less than 6,500 feet. When stratocumulus clouds fill the whole sky, they have a wavy look (Plate 13).

Stratus clouds, on the other hand, are low clouds of indefinite shape (Plate 14). They give the sky a hazy appearance. Although a stratus cloud often averages about 1,000 feet in thickness, many times it is only 50 to 100 feet deep. Then it is usually a local cloud, and clear sky can be seen through it as the cloud breaks up.

Farther aloft, based at average elevations from 6,500 to 20,000 feet, are altostratus and altocumulus clouds. Altostratus clouds form a gray or bluish-colored fibrous sheet of variable thickness over the sky (Plates 5 and 16). Sometimes the sheet of clouds is thick and dark, and may occasionally hide the sun or moon. Other times, it is highly translucent. If the sun appears through altostratus clouds, it looks as though it were shining through ground glass, and it never casts shadows.

Altocumulus clouds occur in small, isolated patches, in parallel bands advancing across the sky, or in a layer, composed of flattened globular masses (Plate 6). Altocumulus often appear at different levels of the sky at the same time. They are of several varieties, and are frequently associated with other kinds of clouds. In summer, they may be seen more at night than in the daytime. When altocumulus pass in front of the sun or moon, a corona, consisting of *small* colored rainbow-like rings with red outside and blue inside, may be observed. If the corona becomes smaller within a short period of time, and other weather conditions are favorable, a storm is approaching.

Unlike cumulus, stratocumulus, stratus, altostratus, and altocumulus, other types of clouds cannot be manufactured in the bathroom. For some are made solely of ice crystals instead of water droplets. These ice clouds form at average altitudes above 20,000 feet when the moisture in the air condenses at temperatures below freezing. They appear as white, thin, delicate clouds, often shaped like mares' tails, mackerel scales, or patches of fleece, and are called *cirrus, cirrocumulus,* or *cirrostratus.*

Cirrus clouds (Plates 2, 7, and 15) are detached clouds of fibrous, feathery appearance. (*Cirrus* is based on the Latin word signifying "curl" or "lock.") The edges of the clouds are ragged and indefinite. Because of their height, cirrus seem to move slowly, although speeds of 100 miles per hour are not uncommon. Before sunrise or after sunset, cirrus may take on magnificent yellow or red colors.

Vhenever they do not merge to form a veil of cirrostratus, ıe cirrus indicate that fair weather will continue.

Cirrostratus clouds (Plate 7) cover the sky with a thin hitish veil. Sometimes the clouds show a fibrous strucure (like altostratus) with disordered threads. Howver, cirrostratus can be easily differentiated from altoratus. For (unlike the case with altostratus) when the ın shines through cirrostratus, it casts definite shadows. nd at any time that cirrostratus clouds pass in front of ither the sun or moon, a halo is produced. It is a *large* ainbow-like circle, with red on the inside and blue on the utside of the colored ring. Because cirrostratus are freuently followed by lower clouds and rain, the halo may e a sign of poorer weather to come.

Cirrocumulus clouds never appear alone in the sky; hey are always associated with cirrostratus or with cirrus Plate 15). Cirrocumulus come in layers or patches, comosed of small white flakes, without shading. The flakes re arranged in groups or lines, or more often in ripples ike those of the sand on the seashore. Because the cloud attern may resemble the arrangement of scales on the ack of a mackerel, it is referred to as a "mackerel sky." If he cirrocumulus give way to cirrostratus and then to ower, thicker clouds, look for wet weather. On the other and, if cirrostratus degenerate into cirrocumulus clouds, he weather will be fair. Hence the saying, "Mackerel sky, oon wet or dry."

It is fortunate that none of these all-ice or all-water louds are able to precipitate markedly. If they were, rain r snow might be pouring from the sky much of the time. Ill-ice and all-water clouds don't usually precipitate beause their ice or water particles aren't large and heavy nough to descend of their own weight without evaporatng immediately. Generally, only when clouds are composed of *both* ice crystals and water droplets can particles e manufactured which are large and heavy enough to decend in the form of rain or snow. (There are a few exeptions to this rule. For example, light drizzle may fall rom an all-water stratus cloud.)

According to theory, as ice crystals are tossed about in he frozen part of an ice-water cloud, they pick up water lroplets on their surface. (For some reason, these water

droplets are unfrozen even though they exist at tempera
tures below freezing.) When the water freezes to the ic
crystals, the mass grows heavy and begins to sink. Whil
it goes downward, it collides with more water droplet
and takes them along, too. The more bumpy the air in th
cloud, the greater the electrical charge on the droplets
and the thicker the cloud, the faster this process is carrie
on. At length, the particle falls to earth as rain or snow
depending on how cold the air is through which it passes
If the particle evaporates into the air before reaching th
ground, long streamers, called *virga,* hang from the cloud

So when precipitation falls, whether over Georgia, Ver
mont, Arizona, or Oregon, it comes from colder region
above. On the hottest of July afternoons, a rainshowe
was once a snowshower! And before the shower came con
densation of moisture and saturation of air. All are prel
udes to spoiling Monday's wash or Saturday's outing.

The three chief ice-water clouds, which bring most rain
and snow, are *altostratus, nimbostratus,* and *cumulonimbus.* Altostratus (Plates 5 and 16) are usually only all-
water clouds. But if the tops build up above the freezing
level and turn to ice, steady rain or snow may result.

Nimbostratus, or "rain clouds," are much like stratus
(Plate 14) except that nimbostratus are so thick that their
tops are icy. Nimbostratus are of a dark, uniformly gray
color. They often develop when a layer of altostratus
grows denser and lower as precipitation falls from it.
The base of a nimbostratus cloud looks ragged and wet;
below this base continuous precipitation frequently falls.
From nimbostratus come the long spring rains.

Cumulonimbus grow out of heavy and swelling cumulus clouds. Heavy and swelling cumulus are thus a transition between cumulus and cumulonimbus clouds. Unlike
cumulus clouds, heavy and swelling cumulus have great
vertical development, and look like giant cauliflowers
(Plate 16). They develop frequently on warm summer
days, or in the strong winds in the rear of a disturbance.
When they form on a hot, calm summer morning, they
mean showers in the afternoon.

If heavy and swelling cumulus clouds sprout tops of icy
cirrus, they become cumulonimbus. The cirrus at the top
of a cumulonimbus cloud spread out in the shape of a

ıantle or anvil. Low, ragged clouds may obscure the ase, which appears dark and threatening. From top to ottom, a cloud may be as much as four miles thick. Cuıulonimbus clouds produce showers of rain, snow, and ometimes hail, and often lightning and thunder as well. 'hey usually generate several layers of altocumulus and tratocumulus clouds; these clouds may extend from the enters of the cumulonimbus over an area of from 20 to oo square miles (Plate 8).

The amount of precipitation that may fall from altotratus, nimbostratus, or cumulonimbus clouds is stupenlous. A rainfall of one inch, for example, over the entire tate of Missouri deposits five billion tons of water on the and. The energy released has been computed to equal oughly that of 140,000 Hiroshima-type atomic bombs! t is little wonder that man-made attempts to cause widepread rainfall by seeding clouds with dry-ice pellets or ilver iodide have often brought unsatisfactory results. n the face of the gigantic forces of the atmosphere, it is urprising that man can influence the weather at all.

Of these three rain makers, the cumulonimbus cloud or thunderstorm is the most fascinating. The Zuñi Inlians, learning to recognize the ivory-towered, anviltopped mass of clouds, discovered that "when the clouds rise in terraces of white, soon will the country of the corn priests be pierced with arrows of rain." Longfellow spoke of them when he wrote.

> The hooded clouds, like friars,
> Tell their beads in drops of rain.

Every year, these storms range over the entire breadth of the United States. According to available weather records, their visits to Tampa, Florida, averaging 94 a year, are more frequent than to any other spot. Santa Fe, New Mexico, is second on the list with 73. Other cities are less frequently visited. The west coast states average only one to four thunderstorms a year (Plate 9).

Lightning and thunder are most bothersome in the months of July, August, and September. During this period, the heaviest thunderstorms strike the central

states. In late winter, on the other hand, the stronge
action is confined to the lower Mississippi valley.

Most storms generate in the daytime. During the sur
mer, however, many of them appear at night in the low
Michigan peninsula, the southwest, and over an extensi
area centered in eastern Nebraska.

Like all clouds, the cumulonimbus is created when
large volume of moist air is cooled to saturation. In th
case, saturation is achieved by violent means. Stron
vertical currents, sometimes 25 to 50 miles per hour (6
miles per hour is the maximum observed speed), hu
moist air upward in turbulent spirals. As the air cool
the main body of the storm boils upward, while decks
stratocumulus and altocumulus are flung outward in a
directions.

If a plane is caught between an updraft and a dow
draft near the center of a cloud, it is in serious dange
Many an accident has resulted when a pilot unwittingl
flew into a thunderstorm masked by clouds and darkne
and then lost control of his ship. Today, fortunately, com
mercial airplanes are equipped with radar devices whic
locate dangerous thunderstorm centers and guide th
planes safely around them.

After the anvil-shaped wedge of cirrus clouds blossom
out on top of the cumulonimbus, rain begins to fall i
dark streaks. Lightning soon begins to flash, because a
the raindrops fall into the strong vertical currents the
are split apart and become electrically charged. When
large charge has collected within a cloud, a spark o
electricity or flash of lightning jumps from cloud to eart
from earth to cloud, or from one part of the cloud to an
other. The length of the stroke is from a fraction of
mile to several miles. Its diameter ranges from one to si
inches. As each flash occurs, the rapid change of electrica
charge in the air causes static on the radio.

The thunder that follows lightning does not spoil milk
Nor will it make hair fall out. It has nothing to do wit
Hendrick Hudson and his crew from the *Halfmoon* play
ing ninepins in the Catskill Mountains. On the con
trary, it is produced by the sudden heating of air sur
rounding the lightning discharge. The rapid expansion
of the warmer air accounts for the sound of thunder.

which can be heard sometimes at a distance of 15 miles. The rumbling of thunder is caused by the crookedness of the path of lightning, by the imposition of the sound of one discharge upon another, and by the reflection of the sound by hills or mountains.

The lightning itself damages or destroys an average of about $18,000,000 worth of property each year. Although it strikes telephone, telegraph, and power lines frequently, it does little harm since they are good conductors of electricity. Installations protected by lightning rods are safe if the rods are properly constructed. But nonconductors, such as brick and stone work or wood, may be demolished. For example, when lightning strikes a tree and forms steam at high temperatures within the pores of the bark, the trunk will split wide open.

Along with its damage to property, lightning manages to kill about 500 people, and injure approximately 1,300 more, every twelve months. Only one-tenth of these casualties occur in districts having a population of more than 2,500 persons, since cases of injury from lightning are rare in modern buildings and houses. So, if you want to play safe, stay indoors during a storm. Keep away from fireplaces, stoves, radiators, bathtubs, and electrical appliances, as well as any other object which is grounded.

When you are caught outside, remember that lightning can and will strike twice in the same place. Avoid isolated trees that have already been struck, especially the pine. Because of its deep, well-grounded roots, a pine tree is to lightning what cheese is to mice. Keep clear of lakes and swimming pools, exposed small sheds and shelters, wire fences, hilltops, and wide open spaces. Find shelter, if possible, in a dense grove of trees or at some low point in the landscape, such as a valley or ravine.

You can tell how near the center of the storm is by counting the seconds between the time you see a flash of lightning and hear the roll of thunder. Every five-second difference indicates a distance of one mile. Consequently, if ten seconds elapse between lightning and thunder, the storm is about two miles away. Since storms generally move at a speed of 25 to 35 miles per hour, you can get an idea of how long you have to find shelter.

Most thunderstorms produce lightning, thunder, and

rain. About one in four hundred is so violent that it generates fair-sized hail as well. The hail forms when exceptionally strong vertical air currents rush upward through the cloud. Large raindrops are carried by these currents into freezing temperatures and frozen into ice. A coating of ice crystals then wraps itself about the particle of ice. As the particle descends, it takes on another coating of water, and then another one of ice. When this process is repeated a number of times, several concentric rings of ice and snow build up over each other, as you can discover by cutting a hailstone in half.

At last the newly formed hailstone becomes so heavy that the rising air currents are unable to hold it aloft, and it drops to earth. The greater the speed of these currents and the greater the moisture content of the cloud, the longer the hailstone will stay in the cloud and the larger it will grow. Estimates indicate that vertical currents must be close to 35 miles per hour to produce a typical hailstone. The largest single stone ever measured in the United States fell at Potter, Nebraska, on July 6, 1928. It weighed $1\frac{1}{2}$ pounds and was 17 inches in circumference and almost $5\frac{1}{2}$ inches in diameter!

While rain descends through the cumulonimbus cloud, it cools the air along the way. This air, being colder than the rest of the air in the cloud, rushes downward also. But because of the variable air currents in the cloud which surge upward and downward like smoke from a cigarette, the rate of fall of colder air and rain is irregular. At one moment, it comes in teaspoonfuls; at another, in bucketfuls as a "cloudburst" with gusty winds up to 50 or 60 miles per hour.

The descending colder air and evaporating rain may cool the air close to the ground as much as 15 to 25 degrees on a hot summer afternoon. If it does not do so to a marked extent, watch for a second thunderstorm to follow.

Cloudbursts often cause the sudden "flash floods" which deluge parts of the western states. The flash floods are formed when heavy rain (usually from thunderstorm cloudbursts over the mountains) falls over a large area and then runs off into a narrow valley or gully. The sudden transformation of a dry gully into a death trap of swiftly moving water gives the impression of a rainfall heavier than has actually occurred. Flash floods may cause great damage, for they frequently wash out bridges and roads and inundate towns.

Flash floods and cloudbursts notwithstanding, you can look forward to thunderstorms without fear. To forecast the exact time of arrival of a thunderstorm over your home or the precise amount of rain it will leave on your garden is out of the question. A general forecast of thunderstorm activity over the state is, however, quite successful, and the average amount of rainfall can be approximated. In summer, when storms are most bothersome, it's wise to close the windows in the morning before leaving home if the temperature is high, the air sticky, the wind from the south, and a few cumulus clouds in the sky. And at any time you hear bursts of static on your radio, you can be sure that a thunderstorm is calling nearby.

Yet the cumulonimbus is but one in the vast array of clouds that can tell you about the weather (see Appendix 2B). The icy cirrus, riding high aloft, announce approaching rain or snow when they thicken and lower. The low-lying stratus, wrapping the hillsides in white, warn of foggy mountain roads. The rounded cumulus, gliding quickly past, inform you of days yet "fair and warmer." In the past, the priests of Delphi were among the first persons to recognize the significance of the changing clouds. When you, too, have become friends with all of them—the polka-dot altocumulus, the haloed cirrostratus, and the puffy cumulus—you'll find that there is more to the weather than is indicated by the thermometer or printed in the newspaper.

6

Pressure and Circulation: Thirty-Four Feet More or Less

> Probable nor'-east to sou'-west winds, varying to the southard and westard and eastard and points between; high and low barometer, sweeping from place to place; probable areas of rain, snow, hail, and drought, succeeded or preceded by earthquakes with thunder and lightning.
>
> —Mark Twain, on New England weather

EARLY in the seventeenth century, Galileo, who a few years earlier had constructed the world's first thermometer, was experimenting with water pumps. Those he made were of simple design. By sucking or forcing air out of the top of the pump tube, the machines created a partial vacuum inside into which the water rose from the bottom. Yet Galileo was puzzled. No matter how well the pumps were built, they could never raise water more than 34 feet. Why, thought Galileo, should they stop at 34 feet?

Torricelli, an Italian physicist and pupil of Galileo's, first provided part of the answer to the question. In 1643, he sealed a glass tube at one end and filled it with mercury (which is about 14 times as heavy as water). Torricelli then inverted the tube in a cup of mercury. When almost 2½ feet of mercury refused to run out of the tube at sea level, Torricelli and his barometer became famous. Never before had anyone designed an instrument to measure air pressure. Never again has anyone designed an instrument more widely used in weather forecasting.

The significance of Torricelli's invention, and of Galileo's question, did not become evident until a few years had passed. In 1647, Blaise Pascal, the renowned French philosopher and mathematician, wrote his brother-in-law, Florin Périer, the following letter:

November 15, 1647

I am taking the liberty of interrupting you in your daily professional labors, and of bothering you with questions of physics, because I know that they provide rest and recreation for your moments of leisure. . . . The question concerns the well-known experiment carried out with a tube containing mercury, first at the foot and then at the top of a mountain, and repeated several times on the same day, in order to ascertain whether the height of the column of mercury is the same or differs in the two cases. . . . For it is certain that at the foot of the mountain the air is much heavier than at the top.*

Nearly a year later, Périer replied:

September 22, 1648

I have at last carried out the experiment which you have so long desired. . . . On top of the Puy-de-Dôme . . . we found that there were 23.2 inches of mercury in the tube, whereas in the cloister gardens the tube showed 26.35 inches. There was thus a difference of 3.15 inches between the levels of the mercury in the tube in the two cases. This filled us with wonder and admiration.*

The experiment, which struck Périer with "wonder and admiration," thus confirmed Pascal's belief that the height to which mercury rose in a barometer was related directly to the amount of air above it.

At sea level, the 2½ feet of mercury in the barometer tube were balanced by the downward force exerted by the weight of air overhead. As Pascal discovered, this force per square inch—or pressure—varies as the amount of air above varies. At elevations above sea level, the total weight of air overhead is less than it is on the surface of the earth, and so the air pressure is lower. As a matter of fact, the air pressure falls off rapidly with height. Mount McKinley in Alaska, only 3⁸⁄₁₀ miles high, stands above more than half of the weight of the atmosphere.

Or, to put it differently, at Times Square and Forty-second Street in New York City, the 500 miles of air overhead exerts an average pressure of 14⁷⁄₁₀ pounds on every square inch of pavement. If you were walking across Times Square, the total pressure of the atmosphere on your body would amount to the weight of eight full-sized

* Reprinted from *A Treasury of the World's Great Letters*, edited by M. Lincoln Schuster. Copyright, 1940, by Simon and Schuster, Inc.

automobiles (between 30,000 and 40,000 pounds). Under such a force, the air close to the earth is greatly compressed and becomes very dense.

But the air farther aloft gets rapidly less dense as there is a lighter and lighter column of air above it to put on the pressure. At less than four miles overhead, the air would exert a total pressure on your body equivalent to the weight of only four automobiles. By the time you reach eighteen miles, 97 per cent of the weight of the atmosphere would lie below you. The air pressure on each square inch of your nose would be less than one-half pound, and the total pressure from head to toe would hardly equal the weight of a Crosley.

In 1935, when the *Explorer II* balloon reached the top of its flight, Captain Stevens glanced at the barometer in the gondola. It showed that the balloon was floating above $^{24}/_{25}$ of the total weight of the atmosphere, and that only $^{1}/_{25}$ remained overhead. And this reading was taken at an altitude of merely 72,395 feet ($13^{7}/_{10}$ miles)!

The average air pressure, consequently, is highest at sea level. There it is able to support an average of about $2^{1}/_{2}$ feet (29.92 inches) of mercury in an evacuated tube. If water is substituted in place of mercury, the air pressure will balance a much higher column, since water is only about $^{1}/_{14}$ as heavy as mercury. Actually, it does balance almost 34 feet of water and no more. So, thanks to the curiosity of Galileo, Torricelli, and Pascal, one of the puzzles of the atmosphere has been explained.

When you drink a milk shake with a straw, you take advantage of this atmospheric pressure. As you suck air from the straw, the liquid in the glass is forced up the straw by the air pressure on the surface. You are refreshed, by courtesy of 500 miles of air aloft, with less effort than it takes to sneeze. If you used a straw 40 feet high, you'd die of thirst before you ever got results!

With the aid of Torricelli's barometer, weathermen have been able to compile records of pressure readings taken in the United States over a period of up to one hundred years. These records show that sea-level pressure readings often do not equal about $2^{1}/_{2}$ feet of mercury, or what is more commonly said, 29.92 inches or 1013.3 millibars. In fact, the records reveal that there is a considerable variation in sea-level pressure readings throughout the United States. Moreover, these variations are a reflection of the unequal heating properties of the mud-pie earth and dishpan-like bodies of water.

Weathermen have found, for instance, that average pressure readings over land are lower in summer than in winter. Over water they are higher in summer than in winter. This seasonal variation in pressure is explained by the mud pie-dishpan fact that land becomes warmer than water does in summer, whereas in winter it becomes colder. The air above land and water likewise reflects these conditions. In summer, for example, the land air is relatively warm and light. It thus exerts an average pressure *less* than that of the cold, heavy air over water. In winter, on the contrary, the land air is relatively cold and heavy. It exerts an average pressure *greater* than that of warm, light air over water.

In summer, consequently, the United States is flanked by two huge mountains of cold, heavy air. These two mountains of air rest over water off the coasts of California and Florida. They are called the "Pacific High" and the "Atlantic High" pressure systems, respectively, because they are areas in which the air pressure is above average or high. In wintertime, when air over water becomes relatively warm and light, the Pacific High and Atlantic High shrink in dimensions. During these months, they are supplemented by systems of low pressure, called the "Aleutian Low" and the "Icelandic Low."

Weathermen have only recently discovered the existence of these pressure systems. Even as recently as Benjamin Franklin's time, the Pacific High and the Icelandic Low were uncharted, and men were as ignorant about the patterns of air circulation as they were about those of air pressure.

Benjamin Franklin was the first person in this country to investigate the relation of wind to weather. He corresponded at length about this matter with his brother in Boston, and with friends living en route between Philadelphia and Boston. Their letters revealed that storms always traveled from Philadelphia to Boston, and never from Boston to Philadelphia. Moreover, this west-east storm movement occurred regardless of the surface wind direction at Philadelphia. Was there, Franklin wondered, an air circulation aloft which might differ from that on the ground? And was this circulation pattern a permanent one?

If Franklin had studied the clouds from day to day, he might have answered his own questions. He would have found that despite the variable wind directions on the ground, there is a permanent stream of air that flows aloft. The direction of this stream of air is most obvious in the eastward movement of the highest clouds.

This steady westerly current of air is called the "Prevailing Westerlies." It sweeps the entire country, with the exception of Florida. Advantage is taken of the west-east "Prevailing Westerlies" air current on nonstop transcontinental west-east speed flights. With the "Prevailing Westerlies" for a *tail* wind, a plane flying eastward can make excellent speed, especially if it flies close to the tropopause where average wind speeds are highest. Commercial airlines, therefore, can travel from St. Louis to New York more quickly on the average than from New York to St. Louis.

During World War II, the Japanese made use of the "Prevailing Westerlies" in a different way. They constucted balloons carrying explosives which they launched in the westerly air current. Knowing that the balloons would drift toward the United States, the Japanese devised timing devices which were supposed to bring the balloons down to earth when they reached this country.

The flow of the "Prevailing Westerlies" close to the ground, however, is variable. It is complicated by numerous ranges of mountains. It is also hampered by local currents of air which arise when one tract of land becomes warmer than another. The hot, ascending air over warmer land is replaced by cooler air from regions having lower temperatures. The horizontal motion of air—or wind—created between the vertical currents of air may blow in any direction, whether with or against the "Prevailing Westerlies." And the up-and-down drafts themselves make the air rough and turbulent. In summertime, such vertical currents may extend upward to 10,000 or 15,000 feet and bounce a plane about violently even on a fair day.

When a plane passes through one of these strong vertical air currents, it is suddenly thrust either upward or downward, depending upon the direction of the current. The abruptness of the thrust gives the impression that the plane has hit an "air fountain" or an "air hole." C. E. Peebles, an airmail pilot, relates how he was flying in a clear sky between Dallas and Kansas City many years ago when he hit an "air fountain" and

> . . . the wind began to boil and swirl. Suddenly and without any warning that anything so unusual was going to occur my ship started upward, completely out of control. With the nose pushed down and the engine wide open, the plane continued to go up. I had been flying at about 1,500 feet. Within 60 seconds, according to the clock on the instrument board, my plane was hurled 8,000 feet higher. As the ship went up it spun much as an aeroplane does when going down in a flat spin. . . . The plane was bouncing around, nose up, then down; straining every wire; being buffeted by the giant winds; completely out of control; going up, up, up. At about 10,000 feet altitude I passed out of it, or maybe it passed away from me. Anyway, the air became reasonably calm and I found that control over the plane had returned.

Finally, the "Prevailing Westerlies" sweep migratory high-pressure and low-pressure systems across the United States in an easterly direction. Unlike the Pacific High, Atlantic High, Aleutian Low, or Icelandic Low, these migratory systems are not fixed in location. Instead, they move in the direction of the over-all circulation of air. Storms which form in their depths are, therefore, carried from west to east. Chicago has a cold wave before the frigid air arrives in New York City. Oklahoma City is usually drenched in rain before people in Nashville put up their umbrellas. San Francisco is often hit by bad weather before the mountains around Salt Lake City are obscured by clouds.

The average rate of movement of these high- and low-pressure systems is about 500 miles a day in summer and 700 miles in winter. If, for example, a high-pressure system is centered over Minneapolis, Minnesota, one winter day and moves at an average rate of speed, it may be close to Dayton, Ohio, the next. Thus you can estimate the distance a pressure system will travel in a day by moving it at a speed appropriate for the season of the year. In winter, the average distance will be greatest. In summer, it will be least.

Another method of estimation is to measure how far the system has gone in the last twenty-four hours. Having done this, you can then forecast that it will move the same number of miles in the next twenty-four hours.

These methods, however, are not always dependable, for a pressure system may move at variable speeds. From Saturday to Sunday, a high-pressure mountain of air may go from Kansas City to St. Louis. But on Monday, it may speed up and travel all the way to Central Park in New York City.

The eastward path taken by an individual pressure system across the country—like its speed—is not fixed. The path taken by a preceding pressure system does have some influence on the path of the system that follows it. Yet both high- and low-pressure systems seem to be quite independent. They may wander wherever they please, and might do something out of the ordinary just to antagonize the weather prophet. Like fingerprints, the patterns of the pressure are never the same. For that reason, they make weather forecasting both interesting and exasperating.

As these pressure systems travel across the United States under the influence of the "Prevailing Westerlies," air also circulates *within* them. Around low-pressure systems, for example, the air circulates counterclockwise toward the center. Around high-pressure systems, it circulates clockwise outward from the center. Thus the wind on the east side of a low-pressure system blows from the average direction of southeast (a "southeast" wind). On the east side of a high-pressure system, it is from the northwest (a "northwest" wind).

The map of October 4, 1950 (Map 5), exemplifies these facts. It reveals a large high-pressure system centered near Kansas City. On this map, drawn from observations taken simultaneously throughout the country at 1:30 P.M. E.S.T., the figure beside each station circle denotes current temperature; at Kansas City, the temperature is 52 degrees. A decimal number beneath temperature indicates precipitation in inches during the past six hours; at Chattanooga, within the past six hours 2/100 (.02) inches of precipitation has fallen. Areas over which precipitation has been observed within the past six hours are shaded; the most extensive precipitation within that time came in the Pacific northwest, around Seattle and Portland.

The solid black lines on the map are called "isobars." They are lines which connect points reporting identical barometric sea-level pressure readings. These isobars are labeled in inches of mercury, and in smaller, more convenient units of pressure called millibars. One inch of mercury equals about $33\frac{86}{100}$ millibars; the standard sea-level pressure of one atmosphere is 29.92 inches of mercury, or 1013.3 millibars. On the map of October 4, Seattle, Tampa, Jacksonville, and New York all report a pressure reading of 29.86 inches, or 1011 millibars.

The circulation of air around the high on this map conforms with the usual clockwise pattern. To the east of the center of the high, at Sault Ste. Marie, Chicago, Cincinnati, and Chattanooga, winds are blowing from a northerly direction. To the west of the center, at Roswell and Bismarck, the winds are southerly.

According to a law formulated by a Professor Buys Ballot in 1857, you can usually tell where high and low pressures lie by observing the wind direction. If you stand with your back to the wind, higher pressure will be on your right and lower pressure on your left. (This rule tends to become more reliable as the wind speed increases.)

Suppose, for instance, you were in Chicago on the afternoon of October 4, where the wind was blowing from the northwest at a speed of from 13 to 18 miles per hour (Map 5). Facing southeast with your back to the wind, higher pressure would be on your right toward Kansas City. Lower pressure would be on your left, to the northeast toward Montreal. Because clearing weather is often associated with high-pressure systems, and because pressure areas travel in the current of the "Prevailing Westerlies" from west to east, you might forecast fair weather for Chicago during the time the wind continued to blow from the northwest and the pressure remained high.

Since the times of Galileo and Benjamin Franklin, the science of weather forecasting has come a long way. Nonetheless, meteorologists today still have much to learn concerning the movement of pressure systems. They also need to know more about the way that pressure fluctuates about 34 feet more and less. Once these phenomena can be forecast with increasing preciseness, it will become simpler to predict the corresponding changes in weather (see Appendix 2A). Each high-pressure system, for example, is composed of a certain kind of air with which a definite type of weather is associated. By identifying the different kinds of air, as will be done in the next chapter, we can come one step closer to putting a finger on one of the most elusive villains in the world—the weather.

7

The Principal Air Masses: Moscow, Trinidad, or Winnipeg?

Out of the south cometh the whirlwind; and cold out of the north.

—*Job*

TWELVE days ago, the air you are now breathing may have rushed up the slopes of Mount Fujiyama. Six days before that, it might have drifted through the halls of the Kremlin. Perhaps three and one-half weeks have elapsed since it wandered by Westminster Abbey. Or possibly it may have spent the last fortnight gliding past steaming Guiana jungles and over the waters of the Caribbean, or dodging ice-sheathed mountains in northern Canada. On the other hand, the particle of air may have been trapped in the vault of your downtown bank for a month. Or it may have been shut up in some dark attic along with grandmother's wedding dress and pictures. Without doubt, its history is as lengthy as a pedigree and as fascinating as an adventure story. But whatever other claim it may have to fame, it's important in helping you forecast the trend of the weather.

Has it traveled from Moscow across the Pacific Ocean before reaching your home on Christmas? If so, it brings part of the moisture and warmth of the ocean with it. Does it make the water glasses sweat in summer and bring crowds to the swimming pools and beaches? The air has recently passed over Trinidad after visiting hot equatorial regions farther southward. Did it come from northern Canada by way of Winnipeg? Then you may expect to feel the frosty tang of ice-and-snow-covered lands in its wintry breath. The weather it brings depends upon where the air originally came from, what direction it has taken since it left home, and how long it has been away.

Your particular particle of air didn't make its journey alone from Moscow, Trinidad, or Winnipeg. It probably was accompanied by many other particles of air of about the same temperature and humidity. Such a mass of air particles, having nearly the same properties, is called—appropriately—an air mass. Thus, if Maine and Montana were under a single mass of air, a plane flying at 10,000 feet above sea level over Maine would report temperature and moisture conditions similar to those reported by another plane flying at that altitude above Montana.

The place or home where air particles assemble to form an air mass is called a "source region." These regions, over which the air moves slowly, are flat and vast. In a short time, the air becomes homogenized throughout as it assumes the temperature and moisture properties of the source region. If the source region is far to the north, the air mass is many degrees cooler than one which comes from the tropics. In addition, if the source region is a body of water like the Pacific Ocean, the air mass which forms over it will contain more moisture than it would if formed over land under similar conditions of temperature.

Once an air mass leaves its source-region home, it roams abroad in the guise of a high-pressure system. As it moves, the mountainous mass of air slowly takes on the characteristics of the lands it visits. In winter, when the air mass moves northward over frozen ground, it is cooled. When it moves southward, it is warmed. At the same time, it may soak up moisture from the earth through the process of evaporation, or lose it through precipitation. It has been estimated that an average of 16,000,000 tons of water are picked up by the air every second from soil, vegetation, lakes, rivers, and oceans throughout the world. Likewise, every time one inch of rain falls over an area of one square mile, the air loses more than 72,000 tons of water.

All these processes gradually modify the air mass until it finally loses its characteristic properties of temperature and humidity entirely. Because of the size of the air mass, this process takes many days. By this time, the air usually reaches the Atlantic Coast, or passes over a number of mountain ranges, before it loses its identity. The change may be so complete that the original air mass is converted into an entirely different one.

There are two main types of breeding grounds for air masses, *continental* over land and *maritime* over water. There are also two main addresses, *polar* to the north and *tropical* to the south. The largest source region of continental polar air for the United States is located in the northern portion of this continent, close to the pole. This region extends over an area from southern Canada to the Arctic Circle, and from the Rocky Mountains to Labrador. In winter, when continental polar air forms over the icy Canadian wastes, it becomes intensely cold and dense. Churchill, Manitoba, on the western shore of Hudson Bay and in the midst of this continental polar source region, has never recorded a January temperature above freezing. At times, readings of less than 50 degrees below zero have been made!

Under the clear skies of its source region, the cold, heavy continental polar air builds up into a high-pressure mountain. When the air pressure on the ground rises sufficiently, the continental polar high-pressure system begins to shove its way southeastward. It may overflow southward as far as Texas while it moves eastward toward the Atlantic Ocean. As the cold wave sweeps onward, Weather Bureau forecasts in cities 1,000 miles to the east read, "Much Colder Tomorrow." Newspaper headlines, following the frigid path of the cold wave, feature news of persons frozen by exposure, of automobile accidents on icy streets, and of destruction to livestock and property.

Even though the continental polar air rushes southward and eastward, it may nevertheless have enough momentum to slide southwestward over the Rockies. When it hurdles the Sierras as well and moves into California, trouble follows. Fruit growers frantically tend their smudge pots while housewives as frantically tear down draperies and curtains before they become soiled with soot. Prices of California oranges may rise in Des Moines grocery stores as fast as the air pressure did in Canada.

At the center of the high-pressure mountain of continental polar air are clear skies or high cirrus "mare's tail" clouds and remarkable visibilities. Although the air is dry at the outset, snow flurries may convert the eastern slopes of the Rockies into giant ice-cream cones when the air is forced upward and cooled to saturation. To the north, the continental polar air picks up both heat and moisture as it moves from Canada over the Great Lakes. These lakes are, for the most part, unfrozen in early winter and warm in comparison to the air. Temperatures rise as much as 15 to 20 degrees on the surface. Since the air close to the water is warmed abnormally, it is forced to rise. Clouds form in the moist rising air currents. They soon return their moisture to the ground in storms and blizzards southeast of the Lakes and along the Appalachian Mountains.

The rise in temperature of the continental polar air when it crosses the Great Lakes does have one good effect. It tempers the climate on the eastern shore by moderating the cold mass of air, by preventing unseasonable warm spells in early spring that bring forth buds which would be nipped by late frosts, and by keeping the summers cool. If you've driven along the eastern shore of Lake Michigan, you may have noticed a strip of land about 30 miles deep. Peaches, grapes, and other delicate fruits are grown there in the famous "Fruit Belt" of Michigan. Nowhere else in the eastern United States are they cultivated so far north.

Meanwhile, the high-pressure mountain of continental polar air ranges over much of the country, cracking milk bottles on back porches in St. Louis and compelling people in New Orleans to put on an extra sweater. It makes excellent skiing weather in New England. Its visits even to New Mexico are so frequent that oranges and grapefruit cannot be grown in the valleys. In Florida, a large part of the orange crop can be ruined by an overnight cold snap. (Only Key West, Florida, has never seen frost.) For the rest of the states, continental polar air means that long underwear is dragged out of closets and put on in a hurry, moth balls or no moth balls.

During the summer, the story of continental polar air isn't half so exciting. In Canada, the white-clad fields of winter, over which the air cooled rapidly during short days and long nights, change into green and black garments. The sun-filled days become longer than the nights. The continental polar air, consequently, is many degrees warmer than in winter. It also contains more moisture, since the lakes and other bodies of water which supply it aren't frozen over.

You can expect, therefore, less violent outbreaks, lower pressures, and cloudier skies in the warm months. There is, of course, cumulus cloudiness in the daytime, but thunderstorms which build out of the puffy cumulus appear no more frequently than in other summer air masses. In most cases, the temperature and humidity drop sharply with the arrival of continental polar air. A long heat wave is ended. The use of woolen blankets is encouraged at night, and the Good Humor ice-cream man is temporarily forced out of business. In winter, continental polar air is thrice cursed; in summer, it is thrice blessed.

The air mass that gives the west coast its agreeable climate is maritime polar air. Since it comes from a maritime source region, it is warmer and more moist than continental polar air. Before arriving on the west coast, a high-pressure mountain of maritime polar air leaves Alaska or Siberia as a cold, dry, disagreeable air mass. However, after drinking freely from the warm waters of the Pacific Ocean, it comes on shore with a much different disposition.

In winter, maritime polar air covers the west coast about three-quarters of the time. When a fresh outbreak reaches the shore, it rushes swiftly up the coastal mountains and over into the San Joaquin Valley to the east, accompanied by gusty winds. As it rides up the slopes, the moist air is cooled and saturated. Clouds rise up above the rugged land and drench large sections of the western coast with showers. In many places, this is the only time of year when precipitation falls.

Meanwhile, in the mountains, violent snowstorms tie up automobile traffic through the passes and break power and telephone lines. Tamarack, at an elevation of 8,000 feet in east-central California, had a seasonal snowfall of 884 inches (more than 73 feet of snow, equivalent to 7³⁄₁₀ feet of rain) during the winter of 1906-7. And even southern California may have a "moist overcast" for several days.

As the maritime polar air moves farther east, it's gradually dried out in its passage over the Cascades, Sierras, and Rocky Mountains. It leaves stratocumulus, cumulus, and cumulonimbus clouds along the mountain slopes as tokens of its visit. But by the time it reaches the Great Plains, skies are seldom filled with clouds. Temperatures are mild with large daily fluctuations.

In summer, maritime polar air is over the west coast all of the time. It brings little rain south of the California-Oregon line. Nevertheless, heavy afternoon thunderstorms do build up on the mountains to the east, and dot the ranges all the way to the Rockies.

Along the coastline of southern California, shallow, low-lying stratus clouds, called "high fog," are an extensive type of nighttime and early-morning cloudiness. Because of the varied elevation of the land, cities close to the Pacific Ocean may have a low overcast of stratus clouds above their rooftops, while the crests of the hills are "socked in" by the "high fog." On the other hand, the cities may be smothered in fog while the hilltops are in the sun. When the stratus clouds settle to the ground, flying conditions around Los Angeles become impossible. Commercial planes are forced to land farther away from the coastline rather than take the chance of landing on an invisible runway.

In addition to the nuisance of "high fog," the Los Angeles basin is frequently bothered by a layer of haze that restricts the visibility. The haze is formed when air filled with salt particles blown off in the ocean spray is trapped in the basin. It may be cleared away by a strong wind, but the haze always returns once the wind has quieted down.

In general, maritime polar air doesn't produce such spectacular weather as continental polar air. For one thing, it flows off a surface of almost constant yearly temperature. Moreover, the moisture it does contain that is capable of producing clouds and precipitation is lost along the western mountains. Much of it, in fact, seems to be lost around Wynoochee Oxbow, Washington, in the foothills south of the Olympic Mountains, where the yearly precipitation averages 150 inches, or 12½ feet.

Maritime polar air thus brings the west coast its mild maritime climate. There are relatively few extremes in temperature during the year, and average winter temperatures are higher than on the east coast. Crops can be grown in the southern part of California with little danger of freezing temperatures, except when continental polar air breaks through the mountain barriers in winter. Whenever bad weather is generated in maritime polar air, however, it makes the headlines. If southern California has rain, the rest of the country soon learns about it!

When high-pressure mountains of maritime or continental polar air enter the southern and eastern part of the United States, neither offers any encouragement to the growth of cotton, tobacco, and sugar cane. Both air masses are relatively dry. Maritime polar air leaves most of its snowstorms and rainstorms along western mountain slopes. Continental polar air, on the other hand, is arid from the start. Both air masses are also chilly. Consequently, if they were the only ones found in the southeast, a Central Steppe of America might cover the region. Cotton, tobacco, and sugar cane would then flourish only in dictionaries.

Fortunately, this is not the case. But where does the rain come from that helps the cypress and dogwood grow? What occasionally floods the mighty rivers that wind through these states, and makes this part of the country a park instead of a wilderness?

The Gulf of Mexico, the Atlantic High centered southeast of Bermuda and extending westward as far as the Great Plains, and the third principal air mass of the

United States—maritime tropical air—bring the answer. The Gulf of Mexico and the adjacent Sargasso and Caribbean seas furnish the supply of moisture. The circulation around the Atlantic High pushes the moisture northward. And the maritime tropical air carries the moisture over land and leaves it in abundant quantities.

Maritime tropical air is both hot and humid. Its source region lies in tropical Caribbean latitudes. From this source region, the air mass has a long trip over warm waters to the United States. In winter, the Caribbean water temperatures range from 70 to 75 degrees, so that maritime tropical air is the warmest air mass to enter the country. It is also the most moist, for it contains an average of fifteen times as much moisture as winter continental polar air does.

After a cold continental polar outbreak has crossed the country in winter, maritime tropical air frequently starts to advance northward over the land that the continental polar air has just vacated. Although the maritime tropical air seldom crosses the Rockies, it often covers many of the central and eastern states. As it first passes over land, it is cooled. Being already near saturation, the air close to the ground soon reaches complete saturation. At night, low stratus clouds and fog quickly develop. The stratus overcast may extend 200 miles inland, and the low bases of the clouds make flying operations hazardous. Like the California stratus, these clouds usually dissolve during the day.

When the air progresses northward, fog and drizzle may occur as the air is cooled further by the frozen land. By the time maritime tropical air reaches New England, it is less warm and moist. And after it passes the coastline and travels over the cold Labrador Current which flows southward past Newfoundland and Labrador, it is cooled even more. Heavy fog, consequently, shrouds this area for more than half of the winter.

Within the air mass itself, thunderstorms are rare, except along the Appalachian Mountains. Here the sharp and rapid lifting of the moist air as it moves up the mountain slopes encourages thunderstorm development.

In winter, the advance of maritime tropical air from the south over the country is normally marked by a welcome temperature rise. Streets are cleaned of ice and snow. The appearance of shirt sleeves and spring dresses is promoted. Confident predictions of an early summer are made. Soon, however, a new continental polar cold wave rushes by. The maritime tropical air is pushed aside and the optimistic weather prophet is compelled to put his winter coat back on and pay off his bets.

A summer outbreak of maritime tropical air is as marked in its appearance as a continental polar cold wave in winter. Temperatures and relative humidities are in the 90's. The nights are hot and unsleepable. The days are humid and unbearable. Thundershowers may bring relief in the afternoon. But they are followed by fog at night and more humidity the next day. The sky is dotted by cumulus clouds, which form in the late morning and disappear after sundown. The winds are southerly. The weather is all heat and stupidity.

During the warmer months, the flow of maritime tropical air onto the southern coast is steady, for the Atlantic High pressure region is strongest in summer. A large outbreak of maritime tropical air frequently occurs when a low-pressure area covers the Mississippi valley. Maritime tropical air is then sucked rapidly northward. Although the air moves off water whose temperature is in the 80's, it is cooler than the land. Stratocumulus clouds cover the shoreline at night and vanish in the daytime. Farther inland, the moist air is warmed by the hot soil and rises upward. Fluffy cumulus clouds fill the sky. Thunderheads boil up in the afternoon and evening; the showers and hail that fall from them soak a wide portion of the countryside.

When the summer maritime tropical air is pushed over the Rockies by the strong circulation of the Atlantic High, thunderstorms become heavy and intense. Along the eastern coast, too, you can see cumulus and cumulonimbus clouds frequently in the daytime. Over the Appalachian Mountains an occasional string or "line" of thunderstorms also develops. After the maritime tropical air passes the New England coastline, it is cooled by the water. As in winter, a fog bank persists out to sea.

Wherever it goes, the maritime tropical "heat wave" has pronounced effects. A score of people who seek relief

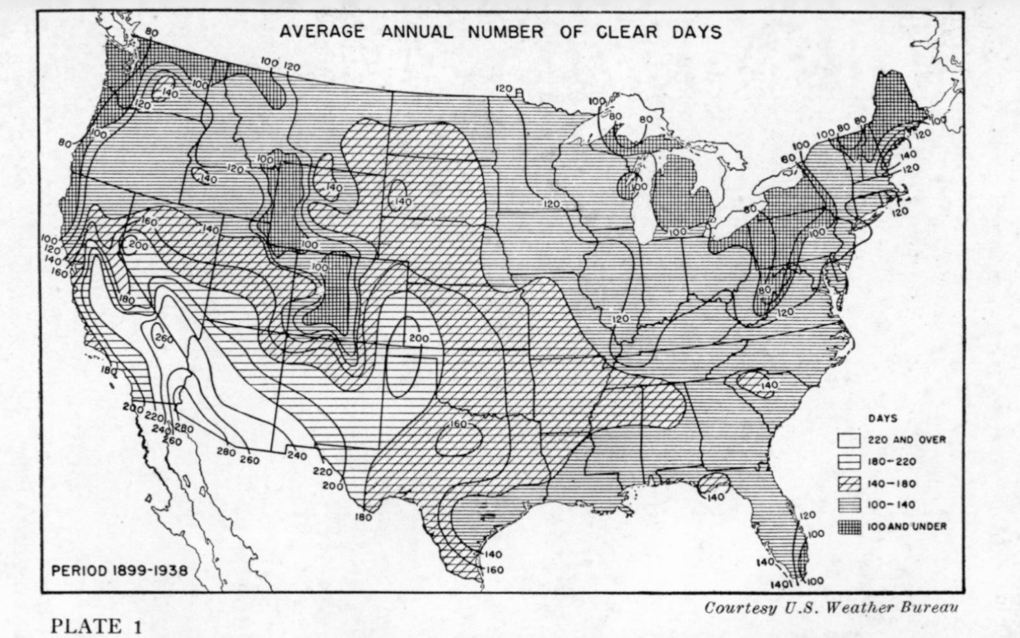

Courtesy U.S. Weather Bureau

PLATE 1

The smallest average number of clear days each year is experienced in portions of Washington, Oregon, Michigan, New York, Pennsylvania, Virginia, West Virginia, Maryland, and Vermont.

Courtesy U.S. Weather Bureau, Agfa-Ansco

PLATE 2

Icy cirrus clouds are above watery cumulus. It will be fine today—if the cirrus do not thicken to rainy altostratus and if the cumulus do not build upward and shower.

Courtesy Standard Oil Company (N. J.)

PLATE 3

Watery cumulus clouds over the Teton Ridge in Jackson Hole, Wyoming. As can be seen, cumulus appear white, shaded, or dark, according to their reference to the sun.

Courtesy Standard Oil Company (N. J.)

PLATE 4

Near Jackson Hole, Wyoming, watery stratocumulus clouds are forming (at the left) from cumulus. If their bases are above 6,500 feet, the stratocumulus are then called altocumulus.

Courtesy U.S. Weather Bureau, Agfa-Ansco

PLATE 5

Altostratus clouds like these may precede heavy precipitation, especially if they thicken, lower, and darken to the west.

Courtesy U.S. Weather Bureau, J. W. Johnson

PLATE 6

Watery altocumulus clouds above Tacoma, Washington. Altocumulus are often seen near thunderclouds, in advance of and behind extensive storms, and on the lee side of mountains.

Courtesy Standard Oil Company (N. J.)

PLATE 7

Icy cirrostratus clouds and cirrus (center) above grain elevators in South Dakota. As pictured here, cirrostratus frequently give the sky a milky color.

Courtesy U.S. Weather Bureau, Air Service, U.S. Navy

PLATE 8

Cumulonimbus with heavy rain. A cumulonimbus suggests the derivation of the word *cloud,* which in Anglo-Saxon meant "rock" or "hill."

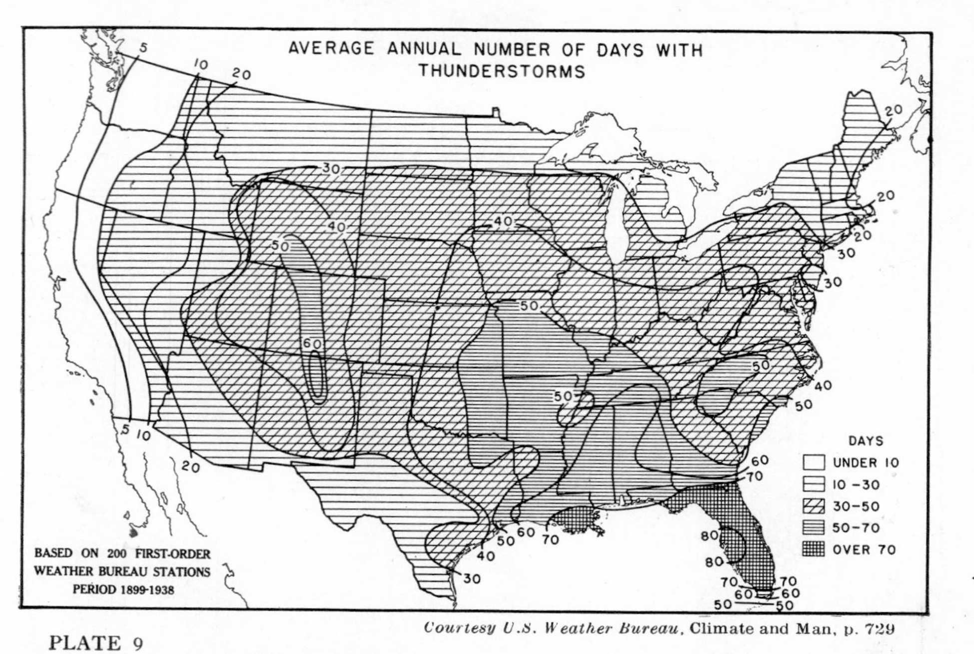

Courtesy U.S. Weather Bureau, Climate and Man, p. 729

PLATE 9

Centers of greatest yearly thunderstorm activity are located near Tampa and Santa Fe. Along the west coast, less than five thunderstorms are usually observed each year.

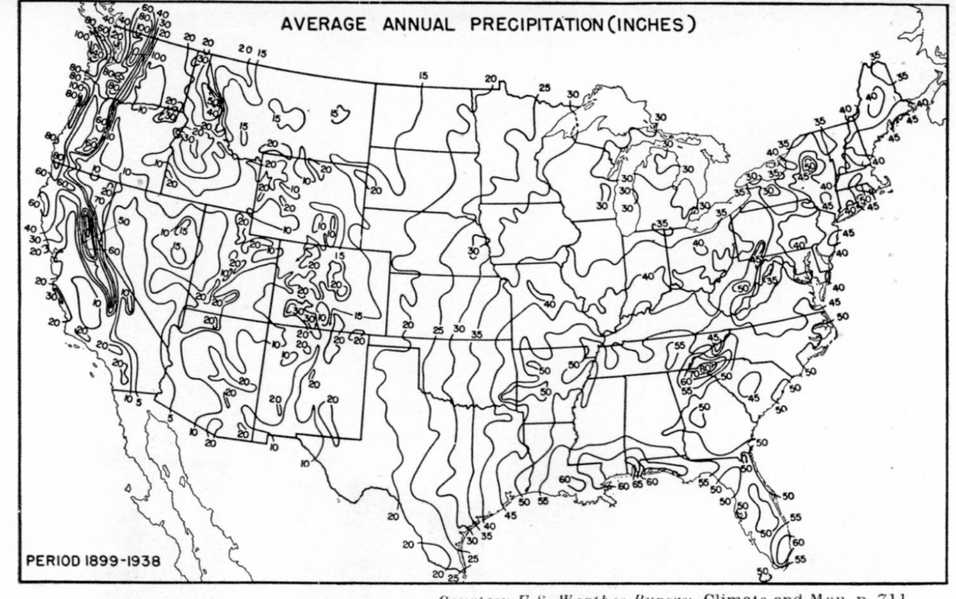

Courtesy U.S. Weather Bureau. Climate and Man, p. 711

PLATE 10

The greatest amounts of precipitation observed each year fall in Washington, Oregon, California, and South Carolina; the smallest amounts in Nevada, Arizona, and California.

Courtesy U.S. Weather Bureau, F. E. Cartrell

PLATE 11

This tornado struck Paris, in western Tennessee, on March 11, 1942, about suppertime. In all, the storm whirled through five counties, and caused heavy damage around Paris.

Courtesy U.S. Weather Bureau

PLATE 12

This picture of noctilucent clouds was taken on July 27, 1909, at about midnight near Dvobak, Norway.

Courtesy Standard Oil Company (N. J.)

PLATE 13

These thick, watery stratocumulus clouds, pictured above U. S. Highway 14 in South Dakota, hide the sun. Oftentimes, a layer is so thin that the sky can be seen through it.

Courtesy Standard Oil Company (N. J.)

PLATE 14

Watery stratus creeping into the Blue Ridge Mountains of Virginia beneath storm clouds. When stratus descend to the ground, they are called fog.

Courtesy U.S. Weather Bureau, Glen Jefferson

PLATE 15

Icy cirrocumulus and cirrus clouds, with a layer of altocumulus below, at Anchorage, Alaska. Because cirrocumulus seldom stay long in the sky, "A dappled sky, like a painted woman, soon changes its face."

Courtesy Standard Oil Company (N. J.)

PLATE 16

A line of heavy and swelling cumulus clouds (with altostratus above) over Weeks Bayou, Louisiana. Soon the swelling cumulus may develop "anvils" of cirrus clouds and become cumulonimbus.

from its warmth at the beaches are drowned. A trainload of cattle suffocate in an afternoon. A tornado writhes downward and wipes out five city blocks in as many minutes. But the corn in Iowa grows two feet overnight, and prosperity comes again to the iceman and the air-cooled theaters.

Maritime tropical air, then, is a breeder of whirlwinds and perspiration; continental polar air of colds and frost-bite; and maritime polar air of whitecaps for the western mountains as well as publicity for the California Chamber of Commerce. Each air mass has its welcome and unwelcome seasons, for each has a particular kind of seasonal weather associated with it. This weather is a reflection of where the air mass was brought up—whether in Moscow, Trinidad, or Winnipeg—and of its subsequent environment. Air-mass weather is, accordingly, fairly consistent (see Appendix 2C). But, as the next chapter indicates, when two air masses come together, most rules fail. The orderly troposphere, composed chiefly of continental polar, maritime polar, and maritime tropical air, is strewn deep with clouds, precipitation, and confusion.

8

Fronts: the Battle Zones of the Sky

The season's first extensive blizzard blew out of the Rockies and roared across Nebraska and Iowa into the Northern Minnesota border area. . . . Fog spread over large sections of the Midwest Saturday and early yesterday, and a "heat wave" developed in the lower Ohio and Great Lakes regions. . . . At least one man froze to death in the blizzard. . . . In the wake of the storm, Lewistown, Mont., reported the mercury at 20 degrees below zero. The heaviest snowfall was ten inches, reported at Livingston, Mont. . . . The advancing cold front was sharply defined, with temperatures ranging more than 50 degrees in Nebraska—from 4 below at Scottsbluff to 49 above at Omaha before the cold winds arrived.

—*New York Times*, Monday, December 12, 1949

DOGS and cats, if properly curbed, are agreeable animals. For one thing, their behavior when separated is dependable—the dog heads for the nearest bone and the cat for the nearest mouse. But once they exchange calling cards, watch out! Past behavior is no guarantee as to what will happen in the future. If an alley cat makes friends with a cocker spaniel one week, she will likely spit in his eye the next.

Air masses are like dogs and cats. When apart, each behaves in a certain manner. Each has its characteristic properties—not of bones and mice—but of warmth and humidity. Winter sub-zero temperatures and skies filled with stratocumulus clouds are typical of continental polar air over Minneapolis, Minnesota. Maritime polar air brings clear, hazy skies on summer afternoons in the vicinity of Los Angeles, California. Humid, thundery days around New Orleans, Louisiana, are frequent with summer maritime tropical air. Although the weather associated with each air mass may not always be pleasant, it is nevertheless dependable. When two dog-and-cat air masses meet and maneuver for position, however, complications develop rapidly.

The boundary which separates a domesticated cocker spaniel from an alley cat is called a fence. But the boundary which separates one air mass from another, and along which one air mass comes in contact with the other, is called a frontal surface.

In many respects, a frontal surface resembles the boundary between cream and milk in a milk bottle. On one side of the boundary is cream; on the other side is milk. The boundary itself is neither pure cream nor pure milk, yet it is distinctly recognizable. The boundary or frontal surface in the air, which separates warmer, lighter air from colder, heavier air, on the contrary, is not visible. Nevertheless, its presence is often announced by lowering pressure, high winds, and an unusual amount of cloudiness and precipitation.

The discovery of the existence of these boundaries was announced by Norwegian meteorologists shortly after the first World War. As the European battle front of the past separated two great armies of men, so a frontal surface in the atmosphere separates two dissimilar mountains of air. Along these battle-zone frontal surfaces, air masses push and shove as each tries to overcome the other.

If you tip a milk bottle, the cream, which i lighter, always stays on top of the milk. Meanwhile, a horizontal boundary remains between them. In the atmosphere, likewise, warm, light air stays above cold, heavy air. At the same time, a frontal surface separates warm air from cold. But this surface, unlike the one dividing milk from cream, which is horizontal, has a considerable vertical slope. The cold air, for instance, may be pushing southeastward at 25 miles per hour while the warm air attempts to shove its way northeastward at 15 miles per hour. The surface along which they meet, consequently, slopes irregularly from the ground to as high as five miles aloft.

The steepness of this vertical slope, or the abruptness of the rise of warm air over cold, determines the type and intensity of cloudiness and precipitation that accompanies the advance of a frontal surface. The more vertica the slope, the more abruptly warm, moist air rises above the cold, heavy air; moreover, the faster the moist ai becomes saturated, as it cools at the rate of about 5½ degrees per 1,000 feet. Clouds, therefore, tend to form more rapidly, and develop vertically to a greater extent as the slope of the frontal surface steepens.

The width of a frontal surface varies from about 5 to 5 miles. Its proportions are fairly well defined. For ex ample, you can always look for a frontal surface in a valley or "trough" of low pressure. Into these valleys flows air from the mountains of higher pressure on either side. These high-pressure regions mark the centers of the air masses separated by the frontal surface.

On newspaper weather maps, you'll notice pencil lines drawn through these "troughs" of low pressure. The lines locate the intersection of sloping frontal surfaces with the ground, and are called "fronts." Upon close examination, you'll find that these fronts separate air masses of high temperature from those of low temperature. They mark the farthest points to which cold or warm air has advanced on the ground. From their position on the ground, the frontal surfaces slope back over the colder air. They may extend 200 to 400 miles away from the surface front while rising aloft several miles.

None of these fronts has a pedigree or label attached to it. Yet each is as clearly identifiable as a Persian kitten or a bottle of Grade "A" milk. Merely by watching the direction a front moves, you can tell whether it is cold, warm, or stationary. (In helping you distinguish the various types of fronts, many newspaper weather maps have "teeth" on the fronts which indicate their directions of movement. The legend on the maps indicates what the teeth signify.)

For example, if a mountain of cold air advances toward a mountain of warm air and, like a snowplow, pushes it backward and upward, the front between them is "cold." On the map of October 2, 1950 (Map 3), a cold front extends from near Winnipeg past Kansas City to Roswell. This cold frontal line marks the farthest points to which the cold air from the northwest has advanced on the ground. It separates an advancing cold continental polar air mass, with temperatures in the 30's to 50's, from a retreating warm maritime tropical air mass, with temperatures in the 70's and 80's.

A simplified cross-section taken between Minneapolis (temperature 47 degrees) and Chicago (temperature 83 degrees) would look like Figure 2.

A "warm front," on the contrary, develops when a mass of warm air advances over and displaces cold air. The map of November 8, 1950 (Map 12), reveals a warm front extending from near Kansas City northeastward past Cincinnati. This warm frontal line divides an ad-

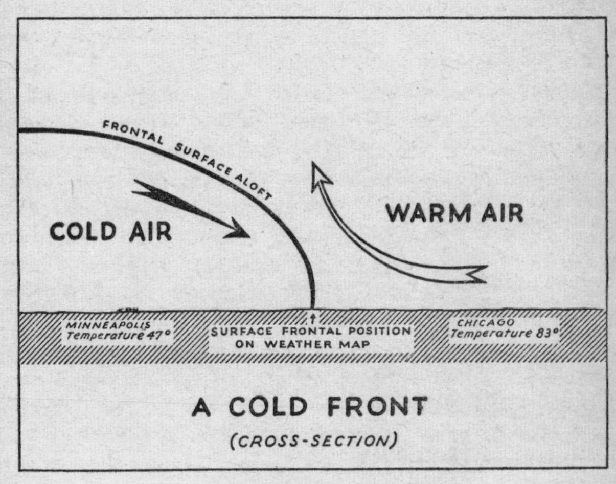

FIGURE 2

vancing warm air mass, with temperatures in the 60's, from a cold air mass to the north, with temperatures in the 40's.

A simplified cross-section taken between Little Rock (temperature 69 degrees) and Chicago (temperature 48 degrees) would look like Figure 3.

In both cases, the *frontal surface* aloft divides the warm air mass from the cold air mass above the ground, just as the *frontal line* does upon the ground. (In both cases, also, the vertical scale is about one hundred times that of the horizontal.)

Generally, cold fronts precede the advance of a mountain of cold air, and warm fronts follow it. The leading slope of the advancing mountain of cold air, therefore, is outlined by a cold frontal surface. The back side or trailing slope is outlined by a warm frontal surface. Since the leading slope of the mountain of cold air is usually steeper than the trailing slope, the weather associated with a cold front often develops more violently than with a warm front.

When a front shows no sign of movement in a definite direction, or wavers back and forth, it is "stationary."

The valley or "trough" of low pressure, in which a

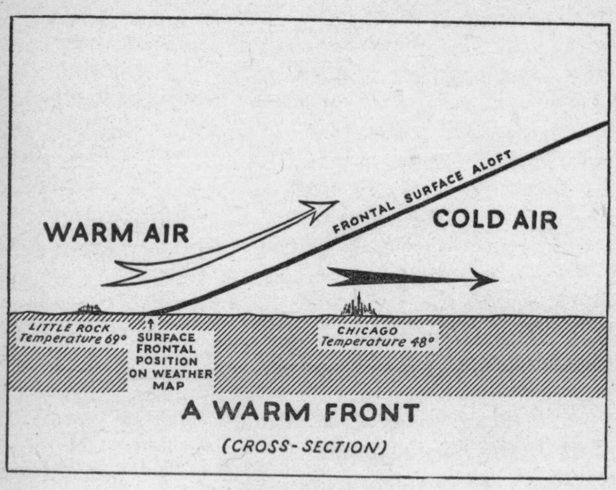

FIGURE 3

frontal surface is located, is often surrounded by strong winds. These winds blow counterclockwise about the frontal surface, as they do around any region of low pressure. In advance of the boundary, the winds are usually from a southerly direction. Behind it, they are northerly. On the map of November 9, 1950 (Map 13), for example, the winds ahead of the cold front at Montreal, Chattanooga, Atlanta, and New Orleans are from the southwest. The winds behind the front at Detroit, Cincinnati, Little Rock, and Galveston are more northerly.

When a front passes you, therefore, the wind itself shifts clockwise. Thus as soon as the cold front on the map of November 9 passes Montreal, Chattanooga, Atlanta, and New Orleans, the winds will shift clockwise from the southwest into a northerly direction. Generally, the more sudden this clockwise wind shift and the greater the change in direction, the more rapid is the change in the weather that accompanies the passage of a frontal surface.

In many cases, if the frontal surface travels fast, the wind shift is marked and the winds are gusty. Moreover, changes in pressure are rapid, the bank of cloudiness associated with the frontal surface is narrow, and the bad weather lasts only a short time. But when the frontal surface moves slowly, winds are light, cloudiness is extensive, and poor weather is persistent.

The strength of a frontal surface (which determines how much weather it can kick up) depends mainly upon the difference in temperature between the air masses it divides. When the temperature contrast is large, the frontal surface is usually strong and beclouded. When the contrast is small, the frontal surface is weak and indistinct. Since temperature contrasts between air masses are greatest in winter, frontal surfaces are generally strongest in that time of year.

The strength of a frontal surface changes constantly as the frontal surface advances over lakes, mountains, valleys, and plains. For as the frontal surface moves, the temperature contrast between the air masses it separates is altered. If, for instance, the warmer air mass loses its heat to snow-covered plains, and if the colder air mass absorbs vast quantities of heat from lakes, then the frontal surface dividing warm air from cold weakens as the temperature contrast between air masses becomes smaller. On the other hand, when fresh amounts of cold and warm air keep pouring into the low-pressure "trough" of the frontal region, then the frontal surface grows stronger as it travels along.

The frontal surfaces that enter the United States generally come from northwestern Canada, from the Pacific Ocean, or from the Gulf of Mexico. Those which form within the United States develop in the southwest or in the Mississippi valley just east of the Rockies. The latter point is a favorite meeting ground for maritime tropical and continental polar air masses. From this spot, they push and shove each other to the east coast and on into the Atlantic Ocean.

Of these three types of frontal surfaces—cold, warm and stationary—the cold front is the most glamorous. It brings the severe blizzards of the winter "cold wave" as well as the welcome cool periods of summer.

When a mountain of continental polar air rushes southeastward from Canada toward the Atlantic Ocean and pushes warmer air (either maritime tropical air from the Gulf of Mexico or maritime polar air from the west coast) out of the way, a cold front forms along the leading edge of the mountain. The easternmost boundary of the cold front between the two air masses usually becomes oriented northeast-southwest. The frontal surface itself often extends from the ground aloft to 10,000 to 15,000 feet. It may slope upward at the rate of 1 mile every 40 to 100 miles along the ground.

This slope (which influences the type and extent of cloudiness associated with the front) changes frequently. As the front surges eastward, the friction caused by varying elevations of land, in addition to smaller obstacles such as trees and buildings, may increase. As a result, the lower several thousand feet of the frontal surface are retarded. The upper portion of the frontal surface, nevertheless, may not be affected. Consequently, the slope of the frontal surface may become more vertical or steeper than ever. At the same time, the cloudiness along the frontal surface may become more pronounced.

Cold fronts move at varying speeds, generally from about 20 to 30 miles per hour (about 500 to 700 miles a day). Thus in twenty-four hours, a strong cold front can leave snow on the rooftops in Omaha, Nebraska, and coat the streets of St. Louis, Missouri, with ice. As a rule, however, the rate of movement is not uniform along the entire length of the cold front, for the front tends to travel faster in its northern portion than farther south. Sometimes, in fact, the southern portion stops completely. When this occurs, the "tail" becomes a stationary front.

The usual combination of sharp slope and rapid movement of a cold frontal surface causes warm air in advance of the frontal surface to be lifted abruptly. In winter, when continental polar air faces maritime tropical air, this sudden lifting of the warm, moist Gulf air often results in heavy rain and snow and strong, gusty winds from Maine to Texas.

Throughout the winter months, the advance of a cold front southeastward from Canada toward our middle western and eastern states heralds the arrival of a "cold wave." In Indianapolis, Indiana, for example, as the front approaches, the atmospheric pressure usually begins to drop, first slowly and then faster. Altocumulus clouds darken the horizon to the west and northwest of Indianapolis. They are soon obscured by stratocumulus and cumulonimbus clouds, as well as by the fall of snow.

When the cold front passes, the snowfall reaches its maximum intensity. The pressure begins to rise. The wind shifts clockwise from a southerly direction into the west. Ordinarily, shortly after the frontal passage, the clouds change into stratocumulus with a few altocumulus. Within 12 to 24 hours after the first snowfall, the sky clears when the main body of dry continental polar air moves overhead. On the other hand, if the pressure does not rise rapidly and if the wind does not veer sharply at the time of the frontal passage, poor weather may likely persist.

Meanwhile, the front speeds eastward, hurling cloudfuls of snow over the countryside. Snowdrifts close highways. Plane flights are canceled by dangerous flying conditions. Hospitals are filled with pneumonia cases. Coal companies are rushed by urgent calls. Three blankets that barely keep you warm at night, plus a frozen radiator in the car the next morning, are visible evidence that the temperature has fallen overnight many degrees because of the cold wave.

Along the Gulf coast, a strong outbreak of cold continental polar air, with its accompanying high winds and snow along the front, is known as a "Norther." These Northers have chilled the land as far south as the Mexican border. They can freeze a herd of cattle with one blast. The fall of temperature may be extreme. San Antonio, Texas, for example, has an average temperature in the 50's in January. But during one memorable Norther the temperature dropped to 6 degrees.

When a cold front approaches the eastern mountains, it often intensifies. In the first place, the air picks up moisture from the Great Lakes, which it leaves along the Appalachian, Green, White, and Adirondack mountains. In the second place, the mountains add an extra upward thrust to the air masses. Consequently, moist air becomes saturated more easily, and cloudiness and precipitation become more widespread.

But when the crest of the mountain ranges is passed, the strength of the cold front is likely to decrease. For one thing, the air masses separated by the front are warmed as they move downslope and their capacity for moisture is increased. Since there is no additional source of moisture available in the mountains, the amount of possible cloudiness is minimized. Furthermore, much of the moisture the air masses held originally has been lost in the Mississippi valley and along the Appalachian plateau. As a result, the people who live in Washington, D.C., might feel a sharp drop in temperature. But they may have relatively little snow (provided the storm comes from the west).

Thus, in the case of a strong winter cold front, cloudiness and precipitation are ordinarily concentrated in a narrow zone close to the front itself.

In summer, cold fronts aren't so pronounced as in winter. Yet along the most marked fronts separating continental polar and maritime tropical air, thunderstorms may be severe. Nevertheless, the 12-hour fall of temperature from the 90's of the humid maritime tropical air to the 70's of the continental polar air makes up for it all. Some people start looking forward to the cooler months ahead as eagerly as they anticipated the warmth of summer during a December continental polar outbreak when all the water pipes cracked.

The spring-like winter maritime tropical air that brings relief from the frigid breath of the cold waves comes northward behind a warm front. The warm frontal surface itself forms along the trailing slope of a mountain of cold continental polar or maritime polar air, as the tropical air pushes the mountain of colder air eastward, and rises over it. On the ground, the warm front normally assumes a west-east orientation (see your nearest newspaper weather map), but the westernmost portion generally moves faster than that farther to the east. By the time the warm front reaches New England, it is often oriented northwest-southeast.

The warm frontal surface usually has such a mild slope that it ascends only 1 mile aloft in every 50 to 200 miles along the ground. The abrupt lifting of warm, moist air associated with a steep cold front is replaced by a more gradual lifting. As a consequence, cloudiness spreads out far ahead of the surface position of the warm front as it moves northeastward at an average speed of between 10 and 20 miles per hour.

Warm fronts develop most often in the southern part of the United States in conjunction with maritime tropical air from the Gulf of Mexico. Or they may choose a breeding ground farther to the north in the Mississippi valley to the east of the Rockies where cold fronts also are born. In either case, the warm fronts travel northeastward. They generally leave the country through New York State or New England.

The warm frontal surface, extending upward 15,000 to 25,000 feet, is outlined by cirrus clouds. These clouds may drift 500 to 1,000 miles in advance of the surface position of the warm front. First comes the tufted cirrus, or "mares' tails," racing through the sky at speeds of up to 100 miles an hour or more. Cirrocumulus clouds follow closely behind. The cirrocumulus "mackerel sky" was so called by sailors who long ago recognized its significance. The fact that cirrocumulus and cirrus are often harbingers of warm frontal rains and high winds prompted a well-weathered jingle:

> Mackerel scales and mares' tails
> Make lofty ships carry low sails.

In the wake of the cirrus and cirrocumulus stream cirrostratus clouds. If these cirrostratus clouds are observed on a moonlit night, a "ring around the moon" may be seen. Since cirrostratus clouds may also proclaim bad weather, "the moon with a circle brings water in her beak."

As the height of the frontal surface lowers when the

front approaches, the clouds lower too. Altostratus and altocumulus arrive overhead, and soon afterward nimbostratus (or occasionally cumulonimbus) clouds may bring snow or rain.

If rain begins to fall through the frontal surface into colder but above-freezing air beneath, the colder air often becomes saturated, and low stratus clouds and fog develop quickly. The hazards of widespread pre-warm frontal "rain fog" and low clouds are recognized by flyers who avoid them as carefully as they do the turbulent cumulonimbus clouds that may be farther above.

After the front passes, a few showers may fall before the sky clears; the temperature, humidity, and pressure rise. Usually, the changes in the weather aren't so rapid as they are with a cold frontal passage. Most of the cloudiness and precipitation comes before the front passes on the surface.

Although warm fronts, like cold fronts, aren't so active in summer as in winter, you can use the same indices to detect their advance in all seasons. Watch first for cirrus clouds merging into altocumulus or altostratus, falling pressure, and rising winds. Then look for rain or snow, fog and warmer weather shortly thereafter.

At any time that a warm or cold front loses its momentum, it becomes stationary. A stationary front, for instance, often forms along the Rocky Mountains during a cold frontal outbreak. When the cold front is unable to ride over the mountains, it remains stationary along their eastern slopes. Generally, the weather connected with a stationary front of this kind is not so severe as it is with a warm or cold front. Yet either type of cold frontal or warm frontal cloudiness may prevail. And no matter what the weather, someone will complain about it!

Like dogs chasing cats, cold fronts are always running after warm fronts. When a cold front catches up with a warm front and collides with it, an "occlusion" results. In some cases, the cold front noses under the warm front and pushes it aloft to form a "cold" occlusion. In other cases, the cold front rides up the warm frontal surface and forms a "warm" occlusion. In either case, on a weather map an occlusion is a line along which a cold front has

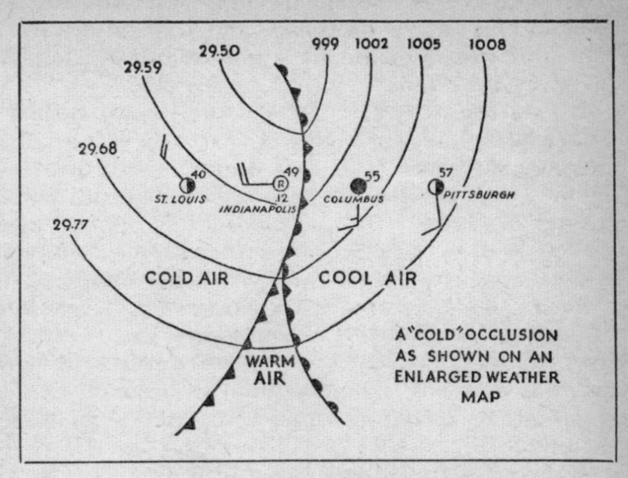

FIGURE 4

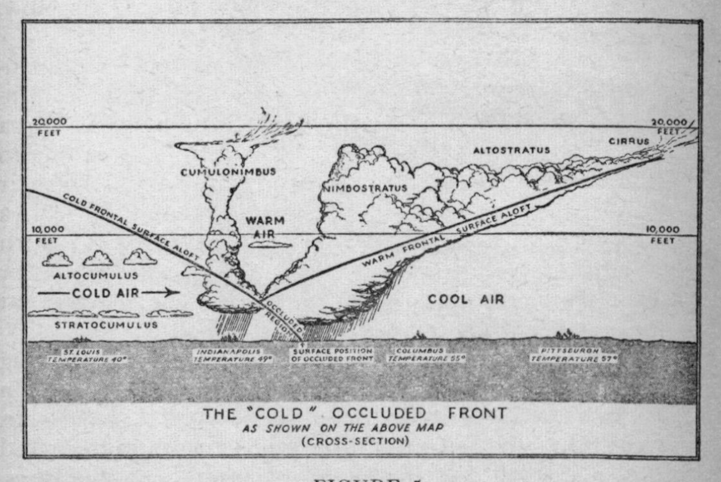

FIGURE 5

overtaken a warm front and lifted the warm air from the earth's surface.

The advance of an occluded front is generally preceded by all the types of clouds which precede a warm front. It is followed by a cloud sequence similar to that following the passage of a cold front. Thus as an occlusion approaches, cirrus clouds first appear in the sky. Then come altocumulus or altostratus, and nimbostratus or cumulonimbus. The pressure falls. The wind increases in speed. Once the point of lowest pressure has passed, the wind shifts. Nimbostratus or cumulonimbus clouds give way to cumulus and altocumulus, which soon begin to clear.

Suppose, for example, that the surface of a "cold" occlusion were located between Indianapolis and Columbus. The line of occlusion of cold and warm fronts might look like Figure 4 on a surface weather map.

This map reveals that cold air lies behind the occlusion, cool air in front of it, and warm air to the south, where the cold and warm fronts have not yet occluded.

A cross-section (Figure 5), taken between Indianapolis and Columbus, would show how the cold front has nosed under the warm front and forced it upward in the region of occlusion.

(The horizontal and vertical scales on this cross-section are not comparable. The horizontal distance between St. Louis and Pittsburgh is roughly 650 miles. The vertical distance between the ground and the cirrus clouds is less than 4 miles.)

This cross-section demonstrates that as the surface frontal position of an occlusion approaches, the cloudiness associated with the occluded front thickens and lowers. For example, Pittsburgh is about 350 miles in advance of the surface position of the occlusion. There the warm frontal surface is far aloft; the clouds are cirrus and cirrostratus. Columbus, however, is only 150 miles from the surface frontal position. Being 200 miles closer than Pittsburgh, Columbus observes thick altostratus lowering to the west. Indianapolis, 50 miles behind the occlusion, reports thundershowers, coming from cumulonimbus along the cold front. St. Louis, 300 miles west of the front, has clearing skies and a falling temperature.

If the occluded front advances toward Pittsburgh at 25

miles per hour and does not change in intensity, rain will begin to fall on Columbus in about three hours. The clouds will thicken and lower to nimbostratus until the occluded front passes in about six hours. At that time, the front will bring thundershowers, and then clearing skies, northerly winds, and temperatures rapidly falling to the 40's.

In about seven hours, rain will start at Pittsburgh, where the same sequence of events will be observed when the surface position of the occlusion approaches and passes. Indianapolis, meanwhile, will find clearing altocumulus-stratocumulus skies and colder temperatures.

Occlusions occur most frequently in the northeastern part of the country, where cold fronts moving out of the northwest overtake warm fronts coming from the south. Such a situation is depicted on the map of November 9, 1950 (Map 13), where the tip of an occlusion is centered north of Sault Ste. Marie. Occlusions also may precede outbreaks of maritime polar air which arrive on the west coast.

These occlusions travel at speeds which depend upon their ages. A young occlusion which is beginning to grow in proportion moves more slowly than an old system which is dying away.

The speed of an occlusion likewise depends upon the depth of the low-pressure area along the line of occlusion. This sector is enveloped by warm and cold frontal clouds, rain or snow, and fog. When the occlusion passes over the country, the pressure at the center may become lower as the cold and warm fronts continue to occlude. In this case, the occlusion will slow up. If the low-pressure area begins to fill, on the other hand, the occlusion will increase in speed. The most practical method of estimating the movement is to forecast that if the low pressure has deepened in the past 12 hours, it will continue to do so in the next period, and move at a slightly reduced rate of speed.

In winter, a limited number of occlusions involving continental polar air from Canada and maritime tropical air from the Gulf of Mexico enter the country. The center of the low-pressure area marking the line of occlusion is usually located far north in Canada. Thus only a small portion of the occlusion trails into the United States. However, continental polar and maritime polar occlusions, or those involving maritime polar and maritime tropical air, are more frequent visitors. During the summer, occlusions of any type are scarcer.

The paths taken by the centers of the low-pressure area marking the line of occlusion are numerous. No two are exactly alike. Yet the low-pressure centers do tend to follow one of three general routes. They may pass across the northern border of the United States (usually with a summer maritime polar-maritime tropical occlusion or with a winter continental polar-maritime tropical storm), often moving from the Dakotas through Michigan and over New England. Or they may follow another track, and swing farther southward from western Canada. In this case, they may pass near Illinois before heading for the northeastern part of the country. During winter, when the Pacific High pressure area retreats from the shores of California, maritime polar occlusions invade the southwest along a third highway. Striking eastward over New Mexico, these occlusions may then drive northeastward through New England by way of Kentucky. Most fair-sized storms, consequently, travel eventually through New England or up the St. Lawrence valley. Few of them (practically none in summer) pass through the southwest. As a result, the weather in New England is diversified while that in the southwest is stable.

The prediction of the exact movement of these fronts, whether occluded, warm, or cold, and of the precise amount of weather that will accompany them, is complicated. Yet you can make satisfactory guesses by basing your predictions of tomorrow's weather on what happened yesterday. For instance, if a cold front moved 550 miles yesterday and increased in intensity, it will do the same today. If an occlusion increased its speed and began to die away yesterday, you can bury it today. If a warm front began to stir up the Mississippi valley yesterday, look for umbrellas and rubbers in West Virginia today.

For the Weather Bureau forecaster who must go into greater detail, the task isn't so simple. Frontal systems never behave exactly alike from day to day. Too often a forecast must be made upon the assumption that the center of the low-pressure area marking the line of occlusion will move north of Detroit, Michigan, and a prediction issued accordingly. Then the system may happen to travel south of Detroit and an entirely unexpected sequence of weather will follow.

It is frequently difficult to forecast for a specific city at a critical spot in the storm's path (and even more so for your back yard) because the difference of only a few miles north or south in the track of the storm means a decided difference in the weather for that city. When a large body of water lies nearby, the situation becomes more complex. Fronts often behave unusually when they approach water. For this reason, the weathermen in Boston, New York, Baltimore, Washington, and New Orleans have a particularly tough job.

Nevertheless, when you locate a front on your newspaper weather map, don't become disturbed. You can find out what kind of front it is by watching how it moves. You can determine its strength by noting how much weather it is causing. You can even predict its movement fairly accurately and outline tomorrow's battle zones of the sky (see Appendix 2E). Of course, fronts aren't so easy to deal with as dogs and cats or milk and cream. On the other hand, you won't be so helpless as the university student who was asked by his professor to draw a cold front. Not having the slightest notion as to what it was, he sketched a picture of an ice box standing in front of a house. He had the "cold" and the "front" correctly. But it was obvious that his ideas were slightly occluded!

9

How to Interpret Newspaper Weather Maps

The cold front extending southwestward from the lower Lakes is expected to reach the Atlantic Coastal Plains by this morning. The high over the Northwest will be centered in the Southern Plains. . . . Colder weather will move to the Atlantic States today, with some snow flurries in the mountains of western Pennsylvania and New York State.

—Weather Bureau forecast for
April 5, 1950

WEATHER conditions over the country on the afternoons of September 30, October 1, 2, 3, and 4, 1950, are shown by Maps 1 through 5, respectively. These newspaper weather maps trace the progress of an outbreak of continental polar air which swept the eastern half of the United States behind a cold front.

On the map of September 30 (Map 1) appear two high-pressure regions. The first is centered in Pennsylvania. The air circulating around this high covers every state east of the Mississippi River. The air is warm maritime tropical air; even north of Florida, its temperature is in the 70's. Cincinnati, Detroit, and Fort Worth report readings as high as 79 degrees. The maritime tropical air moving off the Gulf of Mexico is also fairly moist. Nearly every city in the Mississippi valley and Great Plains observes cloudy skies filled with various types of cumulus. Close to New Orleans and Tampa rain has fallen in the last six hours.

The second high-pressure region is located above Calgary in western Canada. Within this cold continental polar air mass, the cloudiness reported is generally stratocumulus and altocumulus. (Information of this sort is

Explanation of the Symbols on the Map Sequences

Figure beside station circle denotes current temperature. A decimal number beneath temperature indicates precipitation in inches within the past six hours. Shading on the map outlines areas of precipitation within the past six hours.

Cold front: a boundary line between an advancing cold air mass and a retreating warm air mass. As the cold air mass advances (usually southward and eastward), it pushes under the warm air mass like a wedge and forces it upward and backward.

Warm front: a boundary line between an advancing warm air mass and a retreating cold air mass. As the warm air mass advances (usually northward and eastward), it glides over the cold air and pushes it backward.

Occlusion (occluded front): a line along which a cold front has overtaken a warm front and lifted the warm air from the earth's surface.

Stationary front: a boundary line between a warm air mass and a cold air mass which shows little or no movement.

Isobars (solid black lines) are lines connecting points reporting identical barometric sea-level pressure readings. Isobars form pressure patterns controlling air flow. Labels are in inches and millibars.

Winds blow counterclockwise toward the center of low-pressure systems, and clockwise outward from high-pressure areas.

Pressure systems usually move eastward at an average rate of 500 miles a day in summer and 700 miles a day in winter.

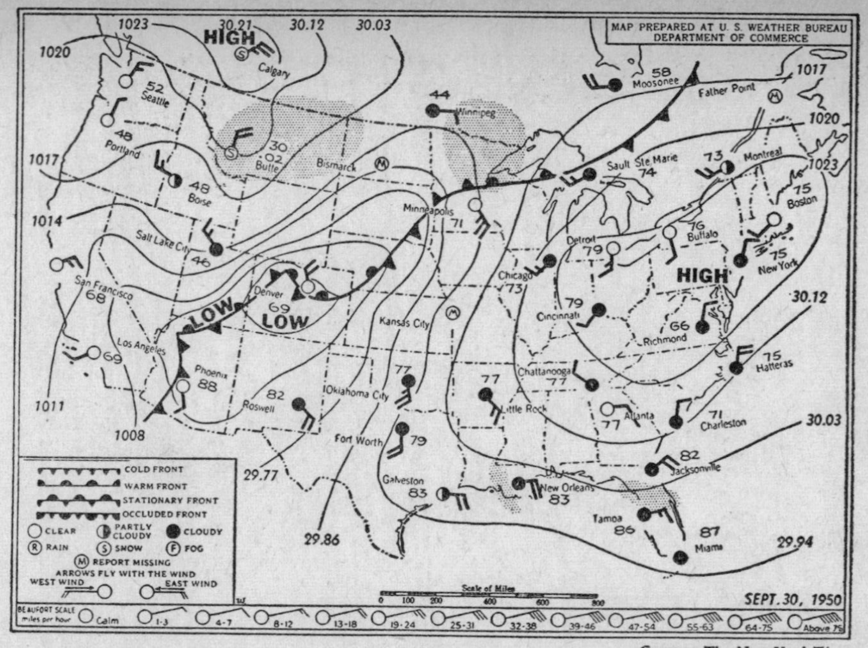

Courtesy The *New York Times*

MAP 1

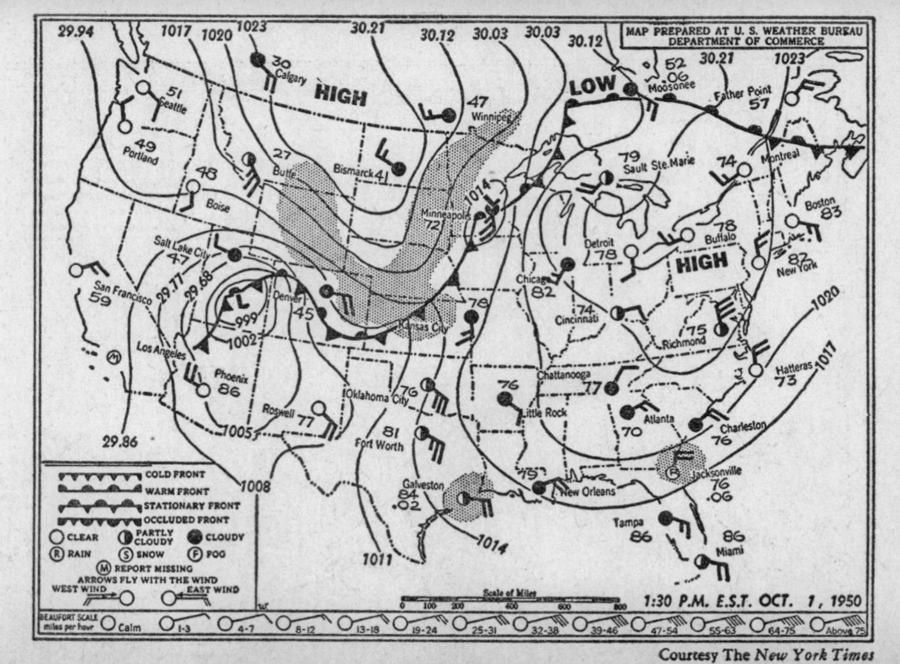

Courtesy The *New York Times*

MAP 2

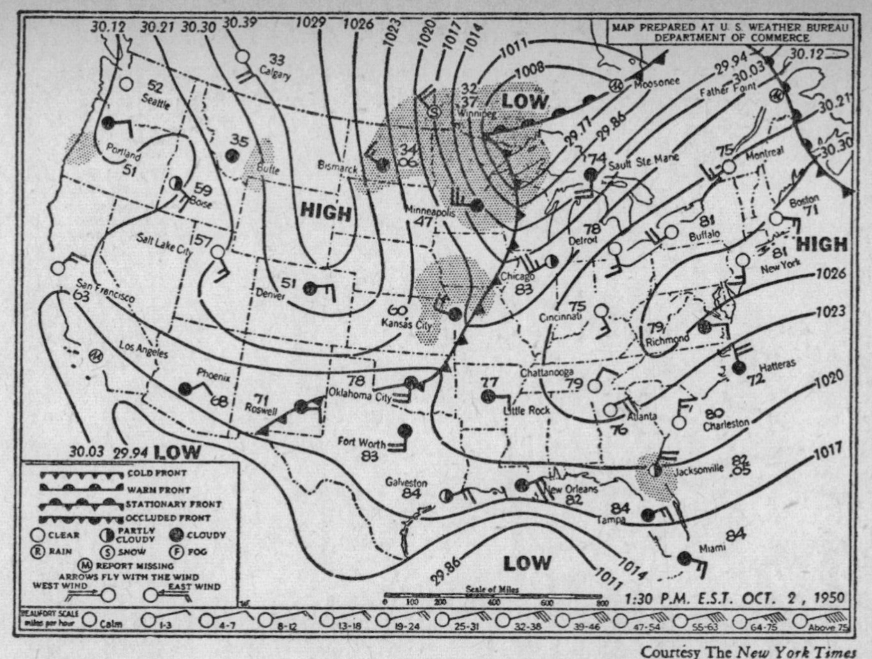

Courtesy The *New York Times*

MAP 3

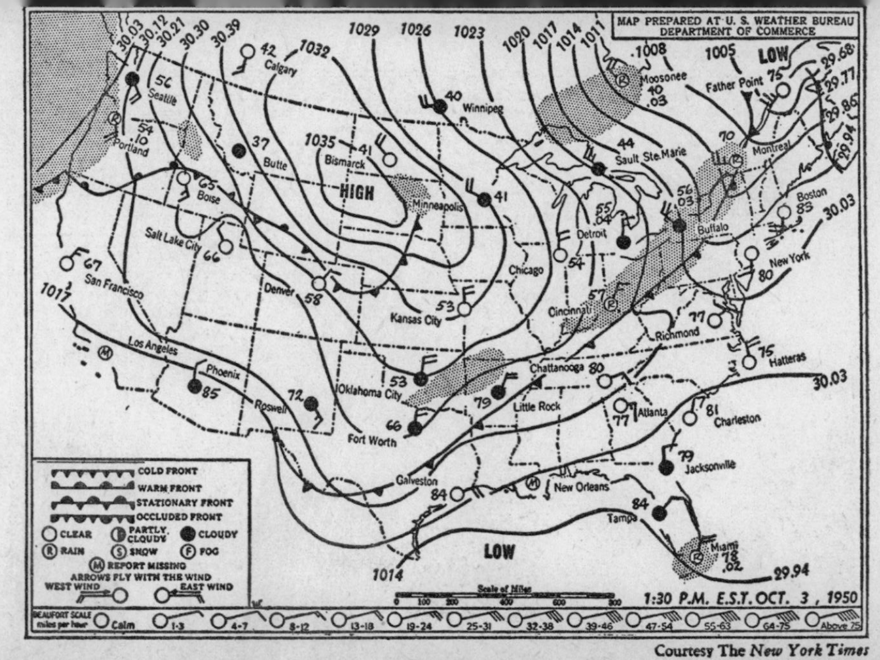

Courtesy The *New York Times*

MAP 4

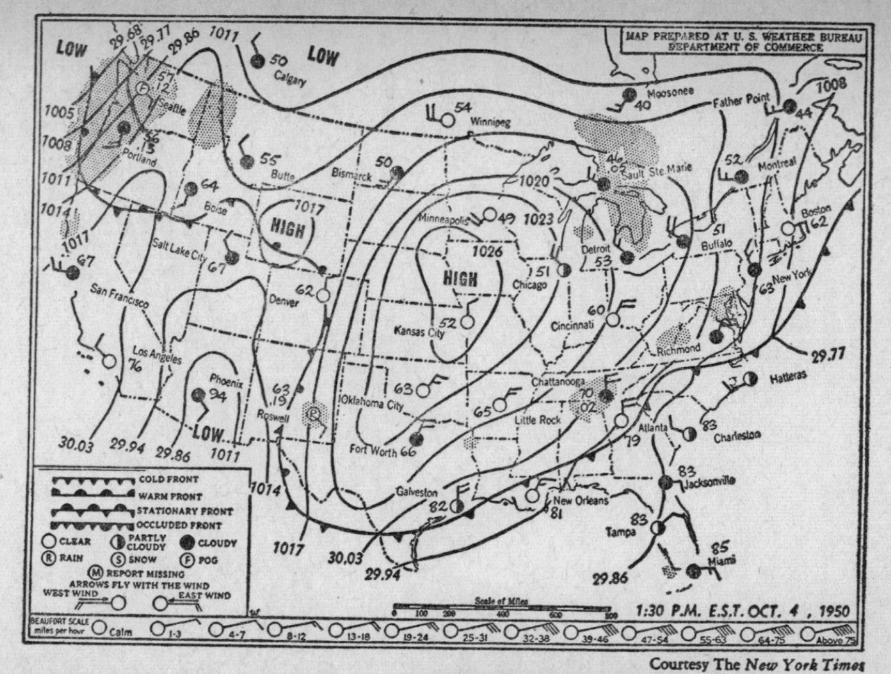

Courtesy The *New York Times*

MAP 5

derived from the Weather Bureau's *Daily Weather Map*.)

In comparison to the warm temperature of the eastern high, the temperature of the Calgary continental polar high is cold. Butte, for example, records a temperature of 30 degrees with snow. Montreal, at nearly the same latitude but under the influence of the warm eastern maritime tropical air, has a temperature of 73 degrees.

The invasion of the cold Canadian continental polar air into the mountains northeast of Butte has resulted in snow flurries within a restricted area, denoted on the map by shading.

The lower portion of the Calgary high is pushing southeastward toward a frontal system which divides the two masses of warm maritime tropical air and cold continental polar air. This frontal system is partly stationary and partly cold. Along it, a low-pressure area is generating in the vicinity of Denver. The fact that the frontal system is partly stationary may account for the absence of marked areas of precipitation around it, even though the front system separates two air masses of decidedly contrasting temperatures. Once it begins to move, however, the system will become more active.

What will happen in the next twenty-four hours?

The Calgary continental polar high will continue shoving its way southeastward past Butte and Bismarck. This is a usual path for a high-pressure region to take when entering the United States from western Canada. If the high-pressure region travels at the average winter rate of 700 miles a day (29 miles per hour), its forward edge should advance behind the frontal system to near Minneapolis and Kansas City. The southerly flow of maritime tropical air over the Great Plains, however, may retard the advance of the cold frontal system. As a consequence, it may not quite reach those cities. Where the warmer, humid maritime tropical air from the south climbs over the cold frontal surface, skies will be clouded with various types of cumulus. Precipitation may fall in limited quantities from local thunderstorms.

The eastern high-pressure region will move slowly toward the Atlantic, as it did in the preceding twenty-four hours. It will continue to circulate warm maritime tropical air over the Great Plains. In this part of the country, temperatures will remain high. Along the Atlantic coast north of Hatteras, temperatures may rise about 5 degrees as the high-pressure region rests overhead and skies clear.

The map of October 1 (Map 2) shows that the eastern high-pressure region stood fast. Temperatures near the northern Atlantic coast rose moderately—at Richmond 9 degrees, at Boston 8 degrees, and at New York City 7 degrees. Two hours after the time of this map, the temperature at New York City reached 88.4 degrees, a record for the date. The *New York Times* reported:

> The mild weather caught the city flat-footed. Heavy coats and woolens brought out by last Sunday's [September 24] near-record low of 43.2 degrees went back into closets and summer apparel was hastily retrieved from bureau drawers and moth bags. Some suburban visitors said buds were popping again in the country. . . . The West did not fare so well. The Associated Press reported that the unseasonable cold which hit the Rocky Mountain area Saturday [September 30] spread yesterday to the Pacific Coast. In some sections of the Rockies the snow was piled as high as sixteen inches.

The continental polar high from Canada moved southeastward to Minneapolis, but did not reach Kansas City. The small advance of the frontal system in this section was accompanied by a narrow but extensive band of precipitation. Between Denver and Butte, the shaded area represents snow which fell as the cold air climbed the slopes of the Rockies.

The Weather Bureau forecast for the next twenty-four hours stated in part:

> Fair, seasonably warm weather will continue over the majority of the eastern states today. But portions of the Middle Atlantic area will have considerable cloudiness with morning drizzle. Showers will be scattered over the Gulf coast, and rain or snow will fall in the North Central States and in the Rockies. It will be colder in most of the Plains States.

This forecast was based on the expectation that the eastern high pressure area would retreat into the Atlantic Ocean as the Calgary continental polar high shoved its way eastward. With the momentum of more and more Canadian continental polar air behind it, the frontal system separating the continental polar and maritime tropical air masses would begin to move eastward more rapidly, accompanied by snow and rain. At the same time, the low-pressure area near Denver would give way to the advance of the Calgary high and dissipate.

By the afternoon of October 2, the chilly Canadian air, moving eastward behind the cold front, has passed over Kansas City and Minneapolis (Map 3). As a result, the temperature at Kansas City is 18 degrees lower, and at Minneapolis, 25 degrees lower. Although the center of the eastern high is in the Atlantic, the circulation around the high continues to pour warm, showery maritime tropical air with cumulus clouds over most of the east. Except for the disappearance of the low-pressure area near Denver, weather conditions in the west have not changed significantly.

How about the weather for October 3?

The upper half of the rapidly accelerating cold front will travel far eastward ahead of the aggressive Calgary continental polar high. If the forward edge of the high-pressure area moves eastward at the average winter rate of speed of 700 miles a day, it would be reasonable to expect the upper half of the cold front to do likewise. This rate of movement would place it near Montreal and Buffalo on October 3. A forecast of cloudy skies, precipitation, colder temperatures, and westerly winds for these two cities would thus be appropriate.

The lower half of the cold front, however, will not move so far. For the push of Canadian air is more eastward than southward. An advance of about 350 miles, therefore, should be expected. This rate of movement would place the front east of Little Rock, and between Fort Worth and Galveston.

If the cold front becomes oriented northeast-southwest between Montreal and Galveston by October 3, the cold continental polar air behind it will cover all of the Great Lakes states. Up to the frontal line, winds will shift into the north or west, temperatures will fall, and precipitation will be widespread. In other parts of the country, weather conditions will remain about the same.

As the map of October 3 (Map 4) indicates, the cold front has surged eastward to the Appalachians. Its passage has brought sharp temperature drops to cities which were formerly under warm maritime tropical air. At Chicago, the temperature fell 29 degrees; at Buffalo and Oklahoma City, 25 degrees; at Cincinnati, 18 degrees; at Fort Worth, 17 degrees. Winds at those cities shifted from southerly to northerly. Within the past six hours, precipitation has fallen from Montreal to Fort Worth along most of the frontal line.

The extensive area of past six-hour precipitation around Seattle and Portland suggests that an occlusion is approaching the west coast. Elsewhere, the Calgary continental polar high, centered southwest of Bismarck, has built up a pressure of 1035 millibars. Some of the cold continental polar air has wandered into the Rockies; a stationary front has developed between Denver and Boise, where the continental polar air was unable to vault the mountains.

One day later, the cold air dominates the entire east (Map 5). The northern portion of the cold front has progressed into the Atlantic Ocean. Temperatures behind it have lowered as the continental polar air poured in overhead. The southern portion of the cold front is trailing through Georgia into the Gulf of Mexico. Maritime polar air covers the west coast, where the Pacific High has moved onshore.

Between September 30 and October 4, therefore, most of the country was cooled off by the advent of the cold continental polar air. The cold front ahead of the Canadian air swept away warm maritime tropical air. Temperatures suddenly fell as the cold front swept by. The largest total drops were observed at Sault Ste. Marie (33 degrees), Chicago (32 degrees), and Buffalo (30 degrees). An inspection of the maximum and minimum temperatures printed in the newspaper on those dates would have revealed even larger total changes than are indicated on these maps.

During this four-day period, rain or snow fell in many parts of the country. Most of the snowfall was observed along the frozen heights of the Rockies, where the cold Canadian air was forced upward, saturated, and beclouded as it tried to climb over the mountains. Some of the rainfall came as a result of thundershowers forming in the maritime tropical air from the Gulf of Mexico. Most of it, however, came as a consequence of the clash of warm and cold air along the cold frontal line. The result of this battle is depicted on the maps by well-defined areas of past six-hour precipitation within those regions through which the cold front moved during that length of time.

The maps of April 3, 4, and 5, 1950 (Maps 6, 7, and 8, respectively), picture the advance of a different outbreak of continental polar air during a spring month. This outbreak traveled much faster than the one of September-October, and caused more rapid breaks in the weather.

The map of April 3 (Map 6) reveals an area of low pressure located near Oklahoma City. Through this low-pressure area runs a partly cold, partly stationary front from west to east. This front separates cold continental polar air to the northwest from warm maritime tropical air to the southeast.

The continental polar air is being swept southward by the action of a growing high-pressure area north of Calgary. (At present, this circulation is assisted by that around a high-pressure region off the west coast which will not affect the picture later on.) The cloudiness reported within the continental polar air mass is chiefly stratocumulus and altocumulus.

The maritime tropical air is being carried northward from the Gulf of Mexico by the Atlantic High, centered east of Hatteras. Within this air mass are stratocumulus and cumulonimbus clouds. On the Gulf coast, New Orleans reports a thundershower. Farther north, Cincinnati, New York City, and Boston record April showers. An extensive band of past six-hour precipitation also shades much of the northeast.

The northernmost advance of the maritime tropical air is marked by a warm front stretching from Detroit past Boston into the Atlantic. Temperatures directly behind the warm front are in the middle 50's; ahead of it they are in the 40's. Cloudiness ahead of the front is cirrus, altostratus, and stratocumulus.

A wide area of past precipitation is connected with both of these frontal systems. This rain has fallen from

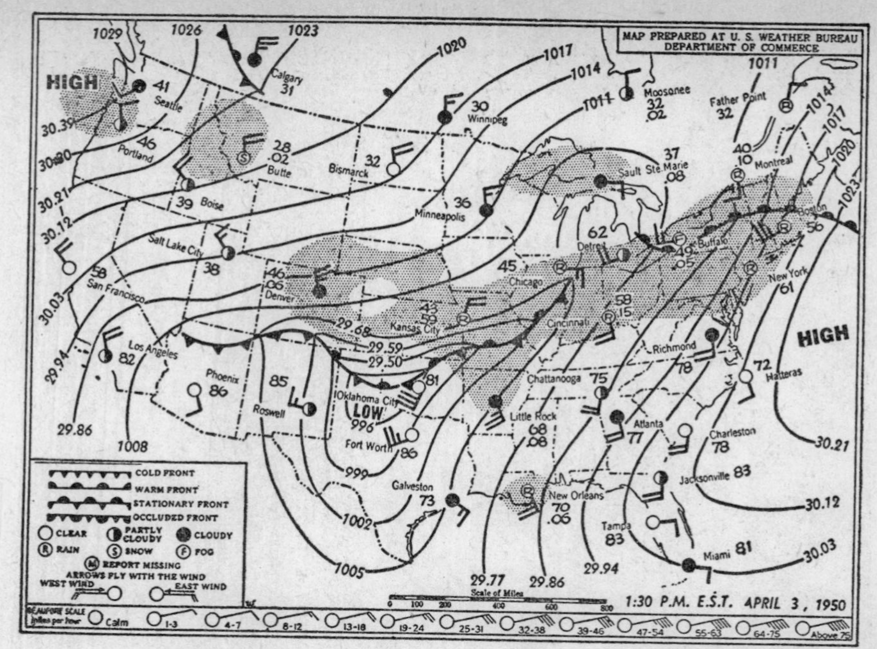

Courtesy The *New York Times*

MAP 6

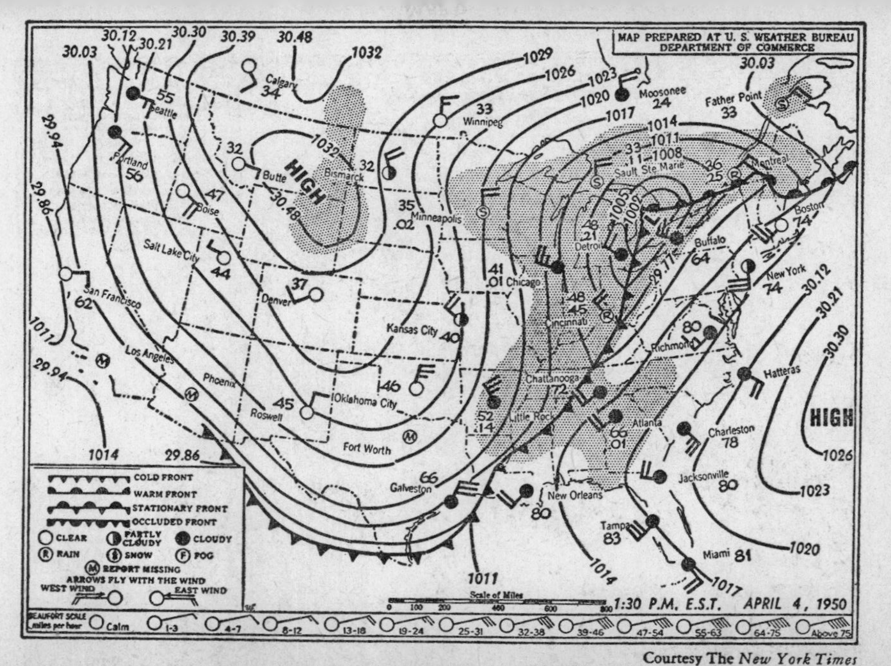

Courtesy The *New York Times*

MAP 7

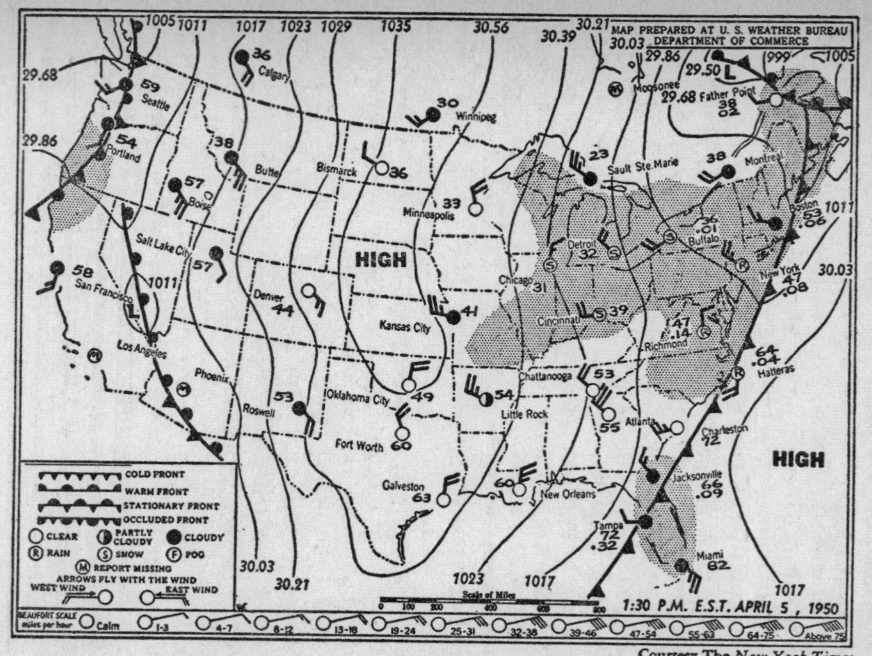

Courtesy The *New York Times*

MAP 8

altostratus and cumulonimbus clouds which surround the regions of frontal activity.

As the northerly winds at Kansas City and Denver suggest, the continental polar high is about to exert a strong southeastward push to the west-east frontal system around Oklahoma City. When the continental polar air slides under the maritime tropical air and pushes it back, the front separating the two air masses will become cold throughout its extent. Within the frontal region, precipitation will be widespread as the moist maritime tropical air is shoved upward and beclouded.

During the next day, a portion of the low-pressure area near Oklahoma City will move up the frontal line into the Great Lakes. Another portion will retreat into Mexico. This course of action is often followed by low-pressure areas which generate in the vicinity of Oklahoma City. Once the low-pressure area moves away, the cold front ahead of the continental polar air should make rapid progress into Texas. In the Mississippi valley, however, it will not move so quickly. For when a portion of the low-pressure area travels up the frontal line and when the cold front intensifies as a result, the rate of movement of the cold front will be slowed up.

This situation calls for a forecast of rapidly falling temperatures, northerly winds, and cumulus-filled skies for southern Texas. In the Mississippi valley, the same forecast will apply, this time with the addition of rainshowers along the frontal line. At the same time, a large area of rain and snow can be expected in the Great Lakes region, where the moisture of the lakes will be added to already disturbed continental polar air.

Along the east coast, temperatures can be expected to increase as the southerly flow of air prevails. Rises will be noted particularly in those cities where rain stops and skies clear. Generally, however, stratocumulus and cumulonimbus cloudiness with local thundershowers will prevail.

When the warm front passes Buffalo, the temperature there should increase to at least that of the warm air (in the middle 50's). If the warm front also travels past Montreal, a similar increase will be felt. But if the front does not reach Montreal, the temperature there will remain between 35 and 40 degrees. On the basis of past twenty-four-hour movement, it is difficult to tell whether the warm front will just pass or just fall short of Montreal. A conservative forecast would be to wait and see.

The map of April 4 (Map 7) reveals how far east the cold front went. Marking the line of farthest advance of cold air, it is situated between Sault Ste. Marie and Galveston. The warm front passed Buffalo, where the temperature has risen to 64 degrees. But the warm air has failed to reach Montreal, which still remains ahead of the warm front. Temperatures in New York City and Boston have shot up under clearing skies and the southerly flow of warm air. In New York City, the sudden rise of 13 degrees to 74 degrees brought a premature feeling of summertime. The *New York Times* noted that "cats dozed in shop windows. The orange-drink stands were rushed, and drugstore counters were lined three deep." Within 24 hours, the cats were to retire to kitchen stoves when the temperature dropped back to 47 degrees as the cold front whipped by.

A temperature rise, amounting to 10 degrees, also occurred in New Orleans. The day before (April 3), when New Orleans observed rain and an overcast sky, the temperature read 70 degrees. However, as soon as the rain stopped and no longer cooled the air, the temperature rose to a more normal figure of 80 degrees on April 4.

Note how sharp a temperature contrast now exists across the cold front. In advance of the front, the maritime tropical air is in the 70's and 80's. Behind it, the temperature of the continental polar air is in the 40's and 50's. The difference augurs a sharp change in the weather to the east, where the cold front will advance swiftly. With it will come April showers with March temperatures.

Unobstructed now by a low-pressure system in the southwest (as on April 3), the cold front can be expected to rush far eastward by April 5. The rapid movement of the front is suggested by the winds of about 20 miles per hour blowing directly toward the front at Detroit, Cincinnati, and Little Rock. These winds indicate that a twenty-four-hour movement of about 500 miles could be expected. This rate of travel would place the front just past Boston, New York City, and Jacksonville. At these three cities, therefore, look for winds shifting to the west, temperatures in the 40's to the north and in the 70's to the south, stratocumulus and altostratus clouds, and showers.

The weather at Buffalo will change again when the cold front overtakes the warm front and occludes with it. According to the usual sequence of events, the tip of the occlusion will go northeastward into Canada toward Father Point. Behind the cold front, continental polar air will cover the nation, bringing with it stratocumulus clouds and woolen blankets.

The Weather Bureau forecast for April 5 thus read in part:

> The cold front extending southward from the lower Lakes is expected to reach the Atlantic Coastal Plains by this morning. The high over the Northwest will be centered in the Southern Plains. . . . Colder weather will move to the Atlantic States today, with some snow flurries in the mountains of western Pennsylvania and New York State.

A glance at the map of April 5 (Map 8) shows that the cold front reached the Atlantic coast. In its wake trailed rain and snow. The deepening low-pressure area along the line of occlusion of the cold and warm fronts traveled into Canada. Over the Great Lakes, the passage of the otherwise dry continental polar air added enough moisture to the air to cause precipitation throughout the region. The center of the outbreak of this continental polar air has descended almost to Kansas City. Along the west coast, an occlusion is coming onshore with fresh maritime polar air.

As the cold front advanced toward Hatteras from April 4 to 5, the pressure there declined from 1020 millibars to about 1014 millibars (the 1014 isobar was not drawn on the April 5 map). This decline was in accordance with the fact that the pressure falls as a frontal system approaches a station.

When a frontal system passes a station, however, the pressure rises. For example, once the cold front traveled by Chattanooga, the pressure rose from a minimum of

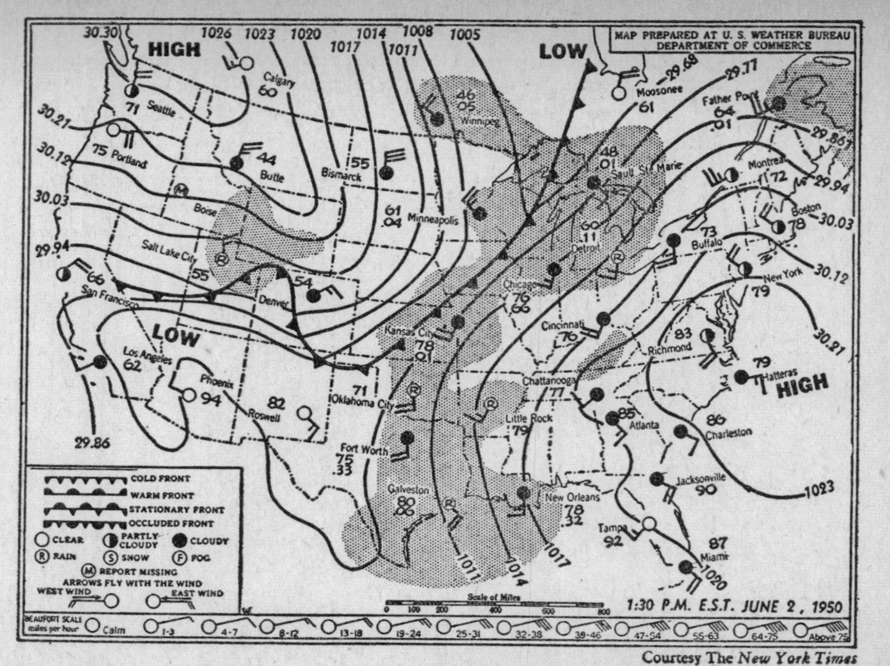

Courtesy The *New York Times*

MAP 9

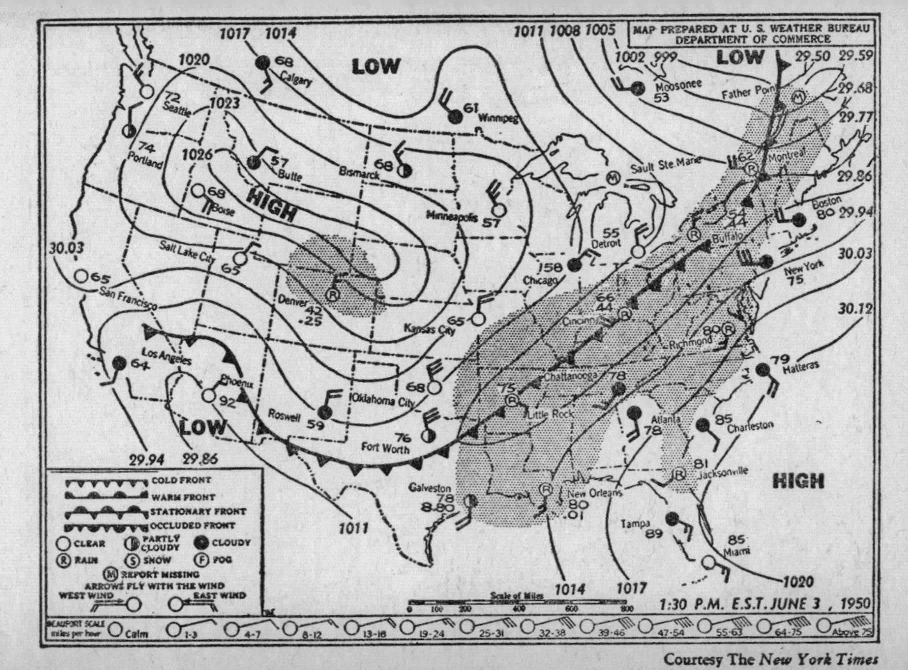

Courtesy The *New York Times*

MAP 10

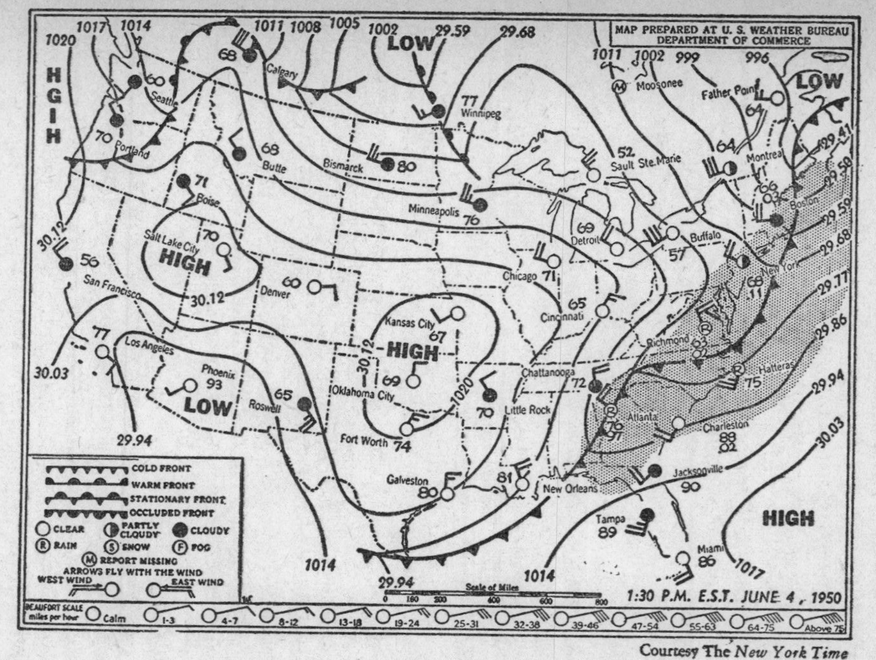

Courtesy The *New York Time*

MAP 11

about 1012 millibars on April 4 to 1023 millibars on April 5. At the same time, the wind shifted from southwest to northwest, the sky cleared, and the temperature dropped from 72 degrees to 53 degrees.

In the three-day interval between April 3 and 5, the southwest portion of the country recorded little cloudiness and no precipitation. The frontal systems which stirred up the weather in the east did not affect the southwest greatly. The strongest effect was a sudden 40-degree drop (from 85 degrees to 45 degrees) at Roswell upon the passage of the "tail" of the continental polar cold front on April 4. Otherwise, the sunny, warm southwest lived up to its publicity.

The southwest is sheltered from sudden changes in the weather for several reasons. For one thing, the "Prevailing Westerlies" tend to blow storms forming near Canada to the east rather than southward. For another, the presence of mountain barriers to the north and east help to steer storms away from the region.

Contrasted to this settled weather situation is that of changeable New England. Through this region passed a warm front and then a cold front in only three days. At Boston, consequently, the temperature jumped from 56 degrees to 74 degrees and back to 53 degrees. Rain and overcast skies were observed on the first and third days, and on the second day a clear sky.

The maps of June 2, 3, and 4, 1950 (Maps 9, 10, and 11, respectively), portray another cold frontal sequence. On June 2 (Map 9), a maritime-polar continental-polar high-pressure region is pointing its finger toward the southeast. A cold front, stretching between the Great Lakes past Kansas City to Denver, marks the farthest points to which the cold air has advanced on the ground. Over the Great Plains, the Atlantic High, centered east of Hatteras, circulates moist warm maritime tropical air northward from the Gulf of Mexico.

Portland, Oregon, reports a clear sky, a temperature of 75 degrees, and a pressure of almost 1023 millibars. Its east wind of 13 to 18 miles per hour conforms to the general pattern of air circulation around the high-pressure region to the north.

Phoenix has clear skies, a temperature of 94 degrees,

and a pressure of less than 1008 millibars. The direction of its light wind has no meaning, since the wind speed is not strong enough to indicate anything significant.

At Oklahoma City, the pressure is 1009 millibars and the temperature 71 degrees. Rain is falling from cumulonimbus clouds in the moist maritime tropical air. The southerly direction of the 13-to-18-mile-per-hour wind anticipates the approach of the cold front to the northwest.

Tampa is clear wih a light southwest wind and a pressure of 1020 millibars.

To the north of the center of the Atlantic High, Boston observes high cirrus clouds, a temperature of 78 degrees, a northwest wind of 13 to 18 miles per hour, and a pressure of about 1018 millibars.

What will the weather be in these cities on June 3?

The eastward advance of the cold front will alter the weather picture at Oklahoma City. If the cold front travels about 700 miles in its northern portion and 350 miles in the south (as it did on the sequence of October 2 and 3), by June 3 it should be oriented northeast-southwest between Montreal and Fort Worth. The cold high-pressure region would then spread far out over the Great Lakes, while penetrating southward as far as Texas.

This cold frontal advance would leave Oklahoma City in cold air some distance behind the front. Such an orientation would call for a forecast of clearing skies, westerly winds, rising pressures, and falling temperatures. The temperature fall may not be too abrupt, however, since the temperature of the cold air is up in the 60's.

Phoenix will not be affected by the advance of the cold front. The forecast is "same as usual."

Portland will continue fair with about the same temperature. The displacement of the continental polar high-pressure area from the northwest to the east may cause the pressure at Portland to fall. It may also give a maritime polar occlusion a chance to slip onshore from the Pacific Ocean into the "trough" of lower pressure created by the eastward displacement of the high-pressure area. This sequence of events occurs from time to time in June. Usually, the occlusion is preceded by high cirrus clouds approaching the coast from the west. Since the main body of the occlusion enters the continent far north in Canada, the occlusion actually causes few marked changes in weather within the United States.

Tampa will note little difference in weather. Some stratocumulus and cumulonimbus cloudiness may develop in the warm maritime tropical air flowing from the Gulf. The pressure may also fall slightly as the surging cold front shoves against the Atlantic High.

Boston will become cloudy with stratocumulus clouds arising in the maritime tropical air. The city also may have occasional rainshowers within the next 24 hours, coming from cumulonimbus.

By June 3 (Map 10), the cold front has swept by Oklahoma City. The pressure there has risen from 1009 millibars to 1014 millibars. The temperature has declined only 3 degrees, from 71 to 68 degrees. (This small temperature decline does not necessarily indicate the cold front was weak. For on June 2, when Oklahoma City was ahead of the front, rain was falling. The precipitation may have caused the temperature to be at least 5 degrees lower than otherwise, and thus have reduced the total temperature decline attributable to the passage of the cold front.)

Phoenix–dry and hot–shows little change.

Portland's pressure has fallen 4 millibars to 1019 millibars. High cirrus clouds are creeping into the western horizon.

The weather situation at Tampa and Boston is about the same.

How about the weather at these cities on June 4?

Oklahoma City will be out of the center of interest; most of the cold front will have passed into the Atlantic Ocean. The high-pressure region gliding over Oklahoma City will bring with it clear skies. The temperature at the city will remain almost constant; the pressure will rise with the advent of the high-pressure region.

At Phoenix, the fair and sunny southwest will remain clear and hot.

Portland's temperature will fall slightly as cooler maritime polar air associated with the occlusion comes onshore. Except for a slight fall in pressure, there will be little net change in the weather.

Tampa will remain warm, humid, and cloudy, with a

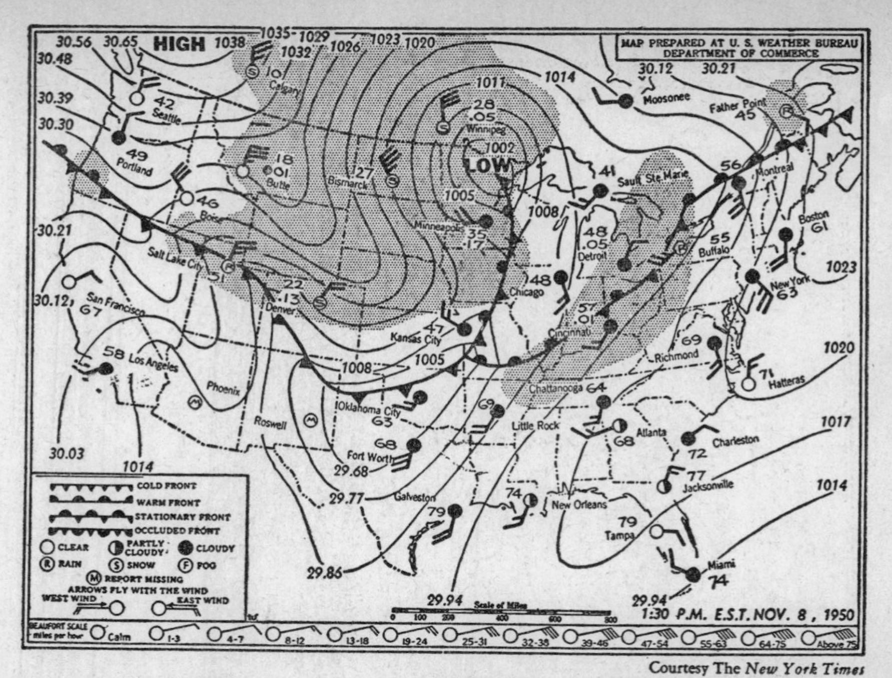

Courtesy The *New York Times*

MAP 12

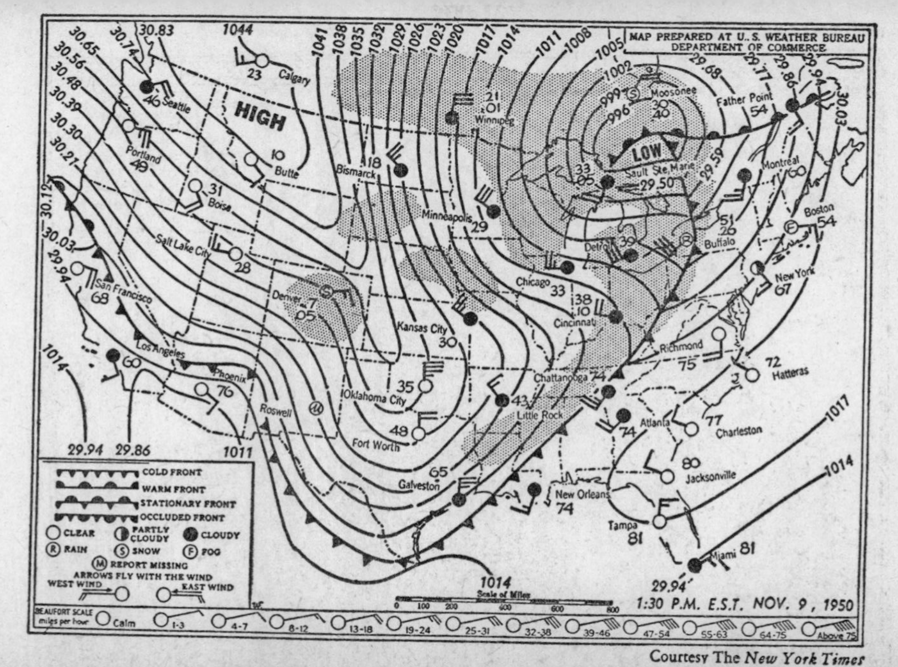

Courtesy The *New York Times*

MAP 13

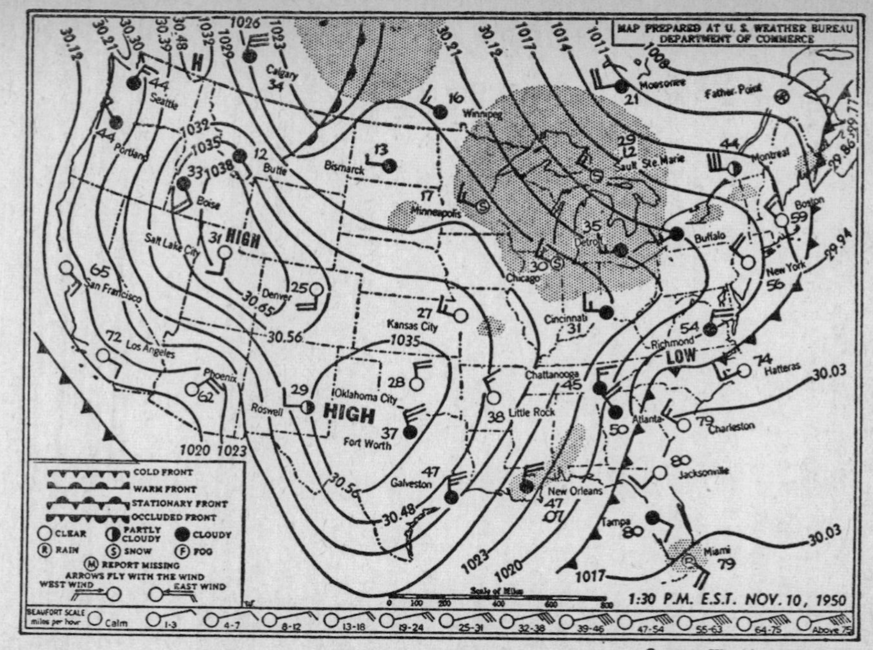

Courtesy The *New York Times*

MAP 14

small pressure decline resulting from the advance of the cold front toward the Atlantic High.

Boston, on the other hand, will find itself behind the cold front. The temperature will drop to the 60's, the pressure will decline and then rise as the frontal area moves by, and showers will develop when the frontal zone moves through the vicinity.

The map of June 4 (Map 11) indicates that the high-pressure area has spread over most of the country. The lower portion of a dry occlusion has entered the west coast without causing any flurries of snow or rain. Except for a slight detour taken in the Carolinas, the eastern cold front has run into the Atlantic Ocean.

The study of these three maps suggests that although the over-all weather picture tends to change rapidly from day to day, the local weather situation at specific points may remain fairly stable. And even at cities located along the frontal highways, the changes were predictable. Of course, the timing and magnitude of such changes is sometimes difficult to predict precisely. But their trend can be readily anticipated from newspaper weather maps.

For practice in using these newspaper weather maps, Maps 12, 13, and 14 are included. They picture the weather sequence from November 8 through 10, 1950, respectively. A comparison of these November maps with those of June (Maps 9, 10, and 11) indicates that temperature contrasts are more noticeable in the colder month of November. In this month, the isobars, or lines of equal pressure, are more closely spaced than in June. This closer spacing reveals more rapid changes in pressure, and hence in temperature, over shorter distances. It also suggests more violent changes in the weather.

Making use of the following condensations of weather summaries prepared by the United States Weather Bureau to accompany each map, the reader may forecast at his own hazard: *For the map of November 8, 1950 (Map 12)*:

The western side of a ridge of high pressure will be the controlling influence over the weather of the Eastern States, while a broad trough of low pressure with centers in the upper Mississippi valley and the Southern Plains will dominate the central third of the nation.

Scattered showers are expected to extend across the Northern States today from the lower lakes region westward to the Northern Plains, where scattered snow flurries are predicted. Showers are also foreseen for the middle Mississippi valley and western Oregon. Light rain is scheduled for Montana and eastern Wyoming with some snow also likely with the showers in the upper lakes region.

Temperatures will remain above normal in the East. Cooler weather is expected in the North Central States westward to the Rockies.

For the map of November 9, 1950 (Map 13):

A trough of low pressure will extend southwestward today from the St. Lawrence valley into the Ohio valley. An intensifying low moving into the lower lakes region and southeasternmost Ontario, with a trough extending further southwestward into the lower Mississippi, will lead into another low in southwest Texas and still another low in Utah.

Much colder weather will overspread the nation today from the Rockies to the Appalachians and from the Great Lakes to the Gulf of Mexico. Precipitation will fall from northern New England southwestward along the Appalachians to the Texas Gulf coast, and in Oregon.

For the map of November 10, 1950 (Map 14):

By this morning the cold air was expected to overspread most of the nation, except the extreme South Atlantic coast on the East and the California coast on the West. The intensely cold high-pressure system was expected to be centered over the Northern Rockies, with the low-pressure cell centered just north of the Great Lakes in southeastern Ontario.

Snow is expected today from the upper Mississippi valley eastward to the lower lake region and sections of the northern Appalachians. Showers are forecast along most of the Eastern Seaboard from Maine to Florida. Except for occasional light rain or snow in the southern Rockies, the weather over most of the western half of the nation will be generally fair.

All of the above map sequences have depicted constantly changing nation-wide weather sequences. Different air masses, separated by various types of fronts, have fought each other incessantly. As a consequence of these atmospheric air battles, the picture of the weather has varied daily on the weather maps. By studying these sequences closely, and then by referring to the maps published in your newspaper, you can predict where the skirmish lines will be tomorrow. Having done this, you needn't be caught with your umbrella down.

10

Fog: One Up on Scrooge

> . . . Where creeping waters ooze,
> Where marshes stagnate, and where rivers wind,
> Cluster the rolling fogs, and swim along
> The dusky-mantled lawn.
>
> —Thomson, *The Seasons*

THE *Christmas Carol* by Charles Dickens starts on a day of "cold, bleak, biting weather." It continues,

> The city clocks had only just gone three; but it was quite dark already—it had not been light all day—and candles were flaring in the windows of the neighboring offices, like ruddy smears upon the palpable brown air. The fog came pouring in at every chink and keyhole, and was so dense without that, although the court was of the narrowest, the houses opposite were mere phantoms. To see the dingy cloud come drooping down, obscuring everything, one might have thought that Nature lived hard by, and was brewing on a large scale.

Dickens was describing a fog in London, a city renowned for fogs so thick that you can get lost in them while crossing the street. But his description would have been just as appropriate for heavy fog in Philadelphia, Kansas City, or Seattle. For, wherever it appears, fog is a "dingy cloud come drooping down, obscuring everything." It often arrives abruptly, stays for a few hours or a few days, and then as abruptly picks up its belongings and disappears.

Note that fog appears in this story on Christmas eve, in the dead of winter. Fog always has its best season over land at this time. Sunny days are short and frosty nights are long (ask Scrooge how long a winter night can be!). The air is cold and sensitive to slight changes in moisture. When its temperature drops a few degrees, or when a little moisture is added to it, the "dingy" stratus cloud begins to droop.

In wintertime, "pea soup" fog comes in the breath of maritime tropical air when the moist air is circulated northward from the warm Gulf of Mexico. As the air travels from the south, it is cooled to saturation in its lowest 1,000 feet by contact with frozen ground or colder water over which it passes. Newspapers soon begin to feature stories like that which appeared in the *New York Times* of Tuesday, December 13, 1949:

> A heavy fog covered New York and vicinity for most of the day yesterday, disrupting airline and ship traffic and slowing movements by automobile and truck. It turned into rain early in the evening. . . .
>
> The fog began to settle in this area on Sunday as a warm air mass moved up from the South. As this air passed over the cold water here, moisture droplets condensed and the fog resulted. It grew denser during the night and by 8 A.M. reduced visibility to 100 yards in the harbor and little more elsewhere.
>
> The same condition prevailed along the Atlantic Coast from Eastport, Maine, to Delaware and as far west as Pittsburgh. . . .

The airports of New York City were closed most of the day on Monday. Airplane landings and take-offs become too hazardous when the visibility (the maximum distance at which large objects can be identified with the naked eye) is reduced almost to zero. Usually the minimum safe visibility is one-half mile. Even then, a commercial airplane landing at 90 miles an hour travels one-half mile in 20 seconds. At this speed, its pilot has only a third of a minute to see an obstruction one-half mile ahead of him and control his plane accordingly.

> The fog also slowed down oceangoing vessels, ferries, and tugs. One ship, the United Fruit liner *Veragua*, came up the harbor under

radar and docked on time. The other ships waited until the mists lifted.

Radar is the only eye that can peer through a thick "pea soup." The human eye is helpless. It is totally blinded by the reflection and diffusion of light which strikes the millions of water droplets suspended in the foggy cloud, each of which acts like a tiny mirror. Radar itself is partially blinded by fog, for the maximum distance it can "see" is considerably shortened when the radar apparatus is pointed straight ahead.

During the summer, "pea soup" is served more frequently over water than over land. Close to the coasts of Labrador and Newfoundland, for example, streams of warm air move off land over the chilly Labrador Current which flows southward from Arctic regions. The bank of rolling fog that arises when the land air is cooled to saturation is often more than one-half mile thick. It is so extensive that it is called the "Grand Banks." In the vicinity of the Aleutian Islands, a similar fog bank persists for most of the year.

When "pea soup" appears along the eastern slopes of the Rockies, it is known as "up slope" or "Cheyenne" fog. This widespread fog forms when moist maritime tropical air from the Gulf is cooled (about 5½ degrees for every 1,000 feet it rises) by being pushed up the foothills of the Rockies. If the wind blows steadily from the east toward the mountains, the "up-slope" fog may blanket every town from Amarillo to Cheyenne a mile deep in white between sunset and midnight.

The very opposite of "up-slope" fog is "steam fog" or "sea smoke." You may see this fog covering rivers and lakes on crisp fall days, or rising out of lowlands "where marshes stagnate" as "swamp mist." "Steam fog" develops when a light wind of unusually cold air passes over warm water. Along the Great Lakes, for instance, as frigid continental polar air from Canada glides over the warm lake waters late in the year, fog may steam upward 100 to 500 feet. On a smaller scale, you can produce "steam fog" in cold weather by placing a pan of hot water outside the window.

A common type of evening fog comes during the summer, after showery afternoon clouds have passed away and the wind has died down. This evening "ground fog" settles first over the countryside. There the ground cools the moist air above it to saturation faster than that in cement-lined cities. Driving from the city after sunset, you'll notice the fog creeping first into valleys and over "dusky-mantled" lawns, and then covering the road and higher ground farther out of town. On such summer nights, therefore, beware of driving into the country. It may be foggier than you think!

When fog is mixed with smoke, the result is a thick, sooty cloud called "smog." Heavy smogs frequent smoke-filled industrial areas when weather conditions favorable for fog last over a period of several days. Ideally, the industrial area should lie "where rivers wind"—in a pocket of a river valley in which the smog can stagnate and thicken. In addition, the air pressure should be above average. For in a high-pressure region, the cold, dense air near the surface prevents any gases, fumes, or moisture near the ground from being carried aloft; the winds are light and cannot blow newly formed fog off the ground; and the clearing skies of the high-pressure region mean that there is little chance of rain which would clean the air and dissipate the smog.

The famous smog in Donora, Pennsylvania, which started October 27, 1948, occurred under these ideal conditions. Located about 25 miles south-southeast of Pittsburgh in the valley of the Monongahela River, the town of about 13,000 persons lay for six days under a large high-pressure region which covered the eastern United States.

On the first of these six days, October 26, the stagnant air near the ground saturated itself with moisture evaporating from the river. Donora was cloaked in fog.

But as this fog was fed with smoke from steamboats, railroad engines, homes, and local steel, wire, and zinc plants, on the next day, October 27, it thickened into smog. The smog spread over an area eight miles wide. It grew so dense that during the daytime the people of Donora "lived in a twilight world."

Two days later, a worker complained, "I couldn't see my hand in front of my face when I went to work. If only it would rain! Why isn't there any wind?" There wasn't any rain and there wasn't any wind because the high-pressure region was still sitting calmly over the eastern United States. Donora was taking the consequences.

The consequences were fatal. Nearly one out of every two citizens became ill as he struggled to breathe the murky air. Investigation showed that the smog was contaminated with several gases known to be irritants. Among those gases was sulphur trioxide, which formed when sulphur dioxide from factory chimneys mixed with the air. These gases, combined with the dusty dampness of the smog, produced an acute irritation of the respiratory tract, and brought death through suffocation to one person after another.

At the end of five days, the death toll stood at 20. It might have risen higher had not the high-pressure region over the east dwindled in size and left for the Atlantic Ocean on the afternoon of October 31. Soon after its departure, a rainstorm swept the smog away. Then Donora went back to the work of living instead of dying.

Fortunately, smog isn't usually so deadly. It seldom persists long enough to do more than deposit a layer of soot on ears, collars, and curtains. Even then, the cumulative effects of seasonal smog in industrial areas may be detrimental to health.

If you live in a smoggy town, you can anticipate dirty curtains and smudgy faces most frequently in October and November, before the changeable weather of winter sets in, when chimneys smoke heavily. At any time that fog is stirred with smoke, however, there will be smog. So you can anticipate its appearance best by looking first for fog.

In winter, then, watch for fog when your newspaper weather maps show rainy maritime tropical air flowing northward from the Gulf of Mexico, or when a large continental polar high-pressure region sits overhead. In summer, you can expect fog in the evening after afternoon thundershowers have cleared up.

Generally, fog appears most readily at night beneath a clear sky, for the ground can cool quickly when there are no clouds overhead. In the daytime, on the other hand, fog often accompanies an overcast sky and drizzle.

So the next time you're caught in a fog (as Scrooge was), you can be sure that it's nothing more than a stratus cloud resting on the ground. You can be certain that it isn't likely to be filled with wailing spirits. If it comes during the day, you know that the sun is on your side. It will be doing its best to heat and dry the air and "burn off" the fog. When it shows up at night, you can expect it to stay until morning unless the wind picks up in speed or drier air moves overhead. But, as the mist begins to thicken, you can be cheered by the knowledge that you're one up on Scrooge. He never knew how or why Nature brews "on a large scale." To him, fog was nothing more than a damp background of a Christmas dream.

11

Winds: Breezes and Blows

Who hath seen the wind?
Neither you nor I.
But when the trees bow down their heads
The wind is passing by.
—Christina Rossetti

IN 1854, during the Crimean War, a windstorm on the Black Sea played an important part in the development of modern weather forecasting. At the time of the storm, the navy of Napoleon III was cruising on the sea and was heavily damaged. In fact, the damage was so severe that Napoleon decided to appoint someone to study weather forecasting and prevent a recurrence of the disaster. He chose the famous French astronomer, Leverrier, for the new job. Leverrier had previously predicted the position of the planet Neptune, and Napoleon reasoned that if Leverrier could forecast the movement of planets millions of miles away, he certainly could handle the weather in his own back yard.

Since Leverrier's time, a greater number of winds have

been discovered than planets. Among those which have been classified are the mountain and valley breeze, the chinook, and the hurricane and tornado winds. No matter whether they carry away the snow on your roof, or the roof on your home, they are essentially alike. Each is a river of air flowing through the atmosphere, traveling from a hill of high pressure to a valley of low pressure. Only when the air temperature and pressure are the same over a section of your state is the air calm. Its quieter periods come in summer and in high-pressure regions.

In summer, whenever you take a vacation in the mountains, you'll run into the "mountain and valley breeze." This is one of the most unusual of all local winds, because in the daytime it actually blows uphill. For under the sun's rays, the temperature of the mountain slopes, and of the air directly above them, rises quickly and soon becomes warmer than the air temperature at the same elevation over nearby valleys. As the warmer air ascends, a stream of valley air breezes up the mountain slopes to form an uphill wind.

This valley breeze, however, isn't so strong as the mountain breeze at night. After sunset, air along the mountains cools more than that over the valleys. Rushing downslope, the colder air may kick up a 50-mile-an-hour wind. An experienced camper will tell you to pamper the mountain and valley breeze by building your fires downhill from camp at night and uphill in the daytime.

Along the eastern slopes of the Rocky Mountains, from southern Colorado to northern Canada, warm "chinook" winds temper the climate in winter. These west winds develop when high- and low-pressure systems are so situated over the continent that a strong downslope airflow is produced which pushes air from the crests of the mountains down their eastern slopes. Air at the mountain crests is thus swept downward and heated about 5½ degrees for every 1,000 feet it falls.

Temperatures at the foot of the mountains rise suddenly. In March, 1900, the thermometer jumped 33 degrees in three minutes one chinook day in Havre, Montana. In December, 1928, a chinook blew through the Mackenzie Valley of Canada. At that time, the temperature in Aklavik, at the mouth of the Mackenzie River, rose in six hours from 5 degrees below zero to 54 degrees above. A few years earlier, as C. F. Sykes reported in "A Bad Winter in Alberta" (*Chambers's Journal*, March, 1927), another chinook struck that western province. He wrote:

The day preceding had been mild with a soft wind blowing out of the east, sure harbinger of bad weather. . . .

The morrow, however, brought an even lower temperature and stronger wind. . . . That night the wind dropped. We awoke to find the temperature standing at fifty-two degrees below. . . . The steam rose in little clouds from the stables, and the smoke from the house-chimneys went up so straight and far . . . that it could easily have been mistaken for a natural cloud. . . . Little pools of mists marked the spots where the cattle stood in huddled bunches, the heat from their bodies combined with their breath hanging over them exactly as mist will gather over a pool on a chilly summer's night. . . .

Our main bunch of horses were on pasture four miles from home when the blizzard struck. In an ordinary winter they would have been able to forage for themselves and grow fat. It was ten days before a man could be spared to see how they were doing. . . .

When the snow finally cleared, dead horses were to be found, with manes and tails eaten off by their starving companions, lying in every sheltered corner, whither they had drifted before the winds, only to find that they had not the strength to turn around and fight their way out through the snow that blew in behind them. . . .

And then, almost as suddenly as it had commenced, the siege was raised; the snow vanished like magic, grass coming up green and fresh as fast as it had disappeared. Horses that had seemed about to die fattened over night. Feed that had gone to famine prices became worthless as it stood rotting in the fields. . . .

In late summer and autumn, the Gulf and eastern coasts are subject to one of the strongest winds of all, the hurricane. In other parts of the world, the storms have various names—typhoons in the China Sea, cyclones in the Indian Ocean, and willy-willies in Australia. Whether willy-willy or hurricane, the results are the same: raging winds, blinding rain, and wholesale destruction.

The hurricanes which strike this country breed either in the North Atlantic in an area east of the lesser Antilles, or in the western Caribbean and Gulf of Mexico. One young storm looks quite like another, and weathermen keep their eyes on all of them.

They can tell something of the strength of the storm from its size. When a big hurricane stirs up, cloudiness may extend over an area up to 600 miles in diameter. The storm becomes doughnut-shaped, with a ring of heavy thunderstorm clouds encircling the center, which is marked by clearing skies, light winds, and the point of lowest pressure. This hole of the doughnut is called the "eye" of the hurricane. It is often about 15 miles wide. Immediately outside the "eye" surge counterclockwise winds of speeds as high as 150 to 200 miles an hour. The winds are frequently so violent that they drive the rain from the thunderstorms in horizontal sheets. The fall of water is phenomenal. A storm in the Philippines in 1911 drenched Baguio, Luzon, with 3⅘ feet of water in twenty-four hours. This record is approached by a twenty-four-hour fall of 2 feet in New Smyrna, Florida, in 1924—considerably more than a drop in the bucket!

When a hurricane nears land, it is preceded by an abnormal rise in the tide, an irregular behavior of the barometer, and high, thickening cirrus clouds. Cirrostratus clouds, with their typical halos, are soon obscured by thicker altostratus and altocumulus. From these signs, the wreck of the *Hesperus* was prophesied:

> Then up and spake an old Sailor,
> Had sailed to the Spanish Main,
> "I pray thee, put into yonder port,
> For I fear a hurricane.
>
> "Last night, the moon had a golden ring,
> And tonight no moon we see!"
> The skipper, he blew a whiff from his pipe,
> And a scornful laugh laughed he.

Shortly thereafter, rain-filled winds reach their maximum speed. The central eye then moves overhead. After it passes, the wind sets in from the opposite direction and the weather comes in reverse sequence.

A hurricane finally dies either by running too far away from the tropical latitudes of its birth, or by traveling over land. Once it hits land, it breaks up fairly rapidly. The uneven landscape hinders the free flow of winds and fails to offer the supply of moisture necessary to keep the storm going.

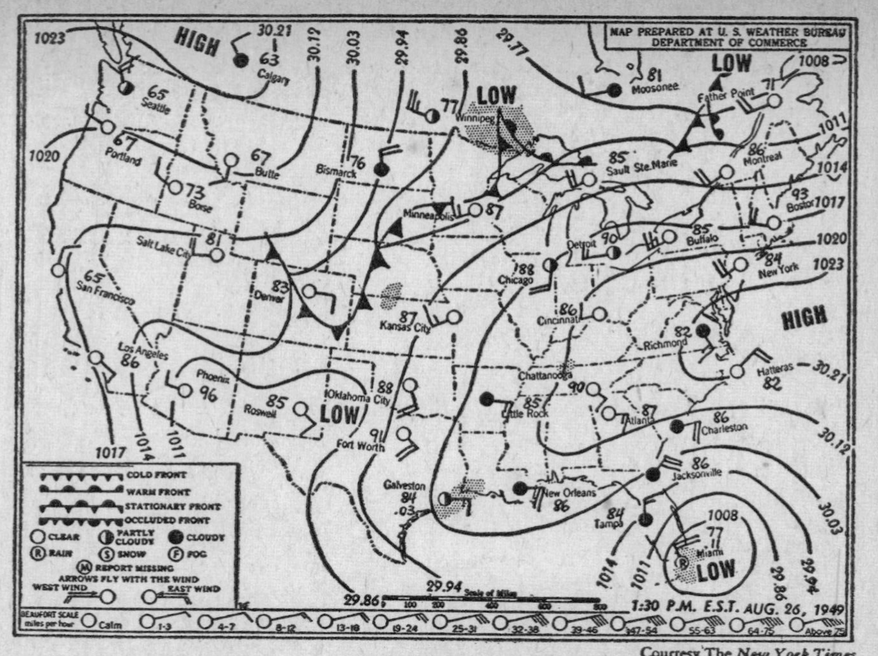

Courtesy The *New York Times*

MAP 15

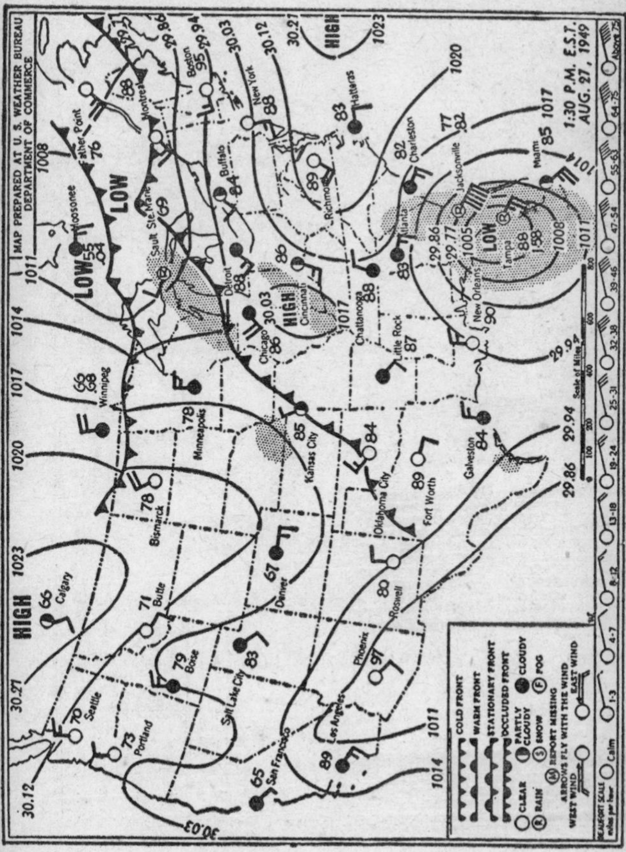

MAP PREPARED AT U. S. WEATHER BUREAU
DEPARTMENT OF COMMERCE

1:30 P.M. E.S.T.
AUG. 27, 1949

COLD FRONT
WARM FRONT
STATIONARY FRONT
OCCLUDED FRONT
CLEAR
PARTLY CLOUDY
CLOUDY
RAIN
SNOW
FOG
REPORT MISSING
ARROWS FLY WITH THE WIND
WEST WIND
EAST WIND

Scale of Miles

BEAUFORT SCALE miles per hour: Calm, 1-3, 4-7, 8-12, 13-18, 19-24, 25-31, 32-38, 39-46, 47-54, 55-63, 64-75, Above 75

Courtesy The *New York Times*

MAP 16

The maps of August 26 and 27, 1949 (Maps 15 and 16, respectively), feature a hurricane which struck savagely at Florida. On August 25, the hurricane was reported to be approximately 480 miles east-southeast of Miami. It was moving west-northwest at about 15 miles per hour. Winds of up to 125 miles per hour were swirling in its depths.

As the map of August 26 (Map 15) shows, the storm, with its typically circular isobars, arrived at the southeast coast of Florida. It passed directly over West Palm Beach, where 125-mile-per-hour winds unroofed homes, smashed plate-glass windows, flattened trees, littered streets and highways with debris, and blew parking-meter heads off their standards. At Jupiter Lighthouse, 18 miles north of West Palm Beach, wind gusts reached 162 miles per hour.

By August 27 (Map 16), the storm had traveled to Tampa. Lashed by howling winds, "grapefruit hurtled through the air like yellow cannon balls." Later that day, the hurricane headed through north-central Florida for Georgia. While weakening slowly, it then wandered past the Carolinas and Virginia to New York, and blew itself out in New England.

As the result of the alert warnings issued by the Weather Bureau, only one person was killed by the storm, and 94 injured. Property losses, however, totaled forty million dollars, with 250 houses destroyed and more than 18,000 homes damaged.

But this hurricane was far surpassed by one that tore through the east coast eleven years earlier. On September 16, 1938, this gigantic hurricane was brewing 1,800 miles east of Cuba. Meteorologists in Florida watched intently as the ugly-looking circular-shaped low-pressure system began to whirl northwest. Two days later, Weather Bureau forecasters at Jacksonville, Florida, noted:

> A tropical disturbance of "dangerous proportions" roared westward over the Atlantic Ocean tonight, its center at 7 P.M. being about 450 miles north of San Juan, Puerto Rico, and about 900 miles east-southeast of Miami.

All the while, as this gigantic atmospheric doughnut moved toward southeastern Florida at 20 miles per hour, the pressure in its "eye" was lowering. The winds, whirling counterclockwise about the center, were strengthening. Cloudiness was spreading. A report from one ship which rode out the storm in the West Indies on September 19 said that the disturbance already covered an area of 240 miles in width.

In Miami, preparations were begun for the storm's arrival. Billboards were removed from their steel frames. Swinging doors were dismantled. Boards were hastily nailed across windows and doors. But the storm changed direction, swerved east of Florida, and charged up the coast.

On the evening of the 20th, the Weather Bureau expected the hurricane to travel "fairly well off the Atlantic seaboard." As a precaution, northeast storm warnings were ordered displayed as far north as Atlantic City, New Jersey. While the warnings were being posted, the hurricane stepped up its speed to 30 miles per hour.

On the fateful 21st of September, the storm unexpectedly swerved again and headed directly toward Long Island and New England at a speed of between 50 and 60 miles per hour. The Weather Bureau frantically warned at 2 P.M.:

> Tropical storm center 12 noon about 75 miles east-southeast of Atlantic City moving rapidly north-northeastward. . . . Storm center will likely pass over Long Island and Connecticut late this afternoon or early tonight attended by shifting gales.

Forty-five minutes later, the hurricane swept Long Island. Huge tidal waves, 30 to 40 feet high, battered the coastline. At Westhampton Beach on the south shore, 144 out of 150 buildings collapsed. On the north of the Island, at Southold, hurricane winds blew the chestnuts off a horse-chestnut tree through the windows of a nearby house, riddling them like machine-gun bullets. The Hudson Tube and ferry slips were flooded, and one foot of water covered the waiting room floors. In the Bronx, winds attained a maximum speed of 78 miles per hour. At the top of the Empire State building (then 1,250 feet high), gusts reached 120 miles per hour. Thousands of fallen trees cut power lines. Electrical and phone service was halted in many parts of Long Island. Meanwhile, the storm roared on.

At 3:40 P.M., the "eye" of the hurricane passed inland east of New Haven. Reports of the storm's rage soon came from all of New England. Providence, Rhode Island, was deluged by a 12-foot tidal wave. Swirling water lapped at second-story windows in the center of town; tens of persons drowned in hotel lobbies. While passengers were "waiting for trains which never came," the roof of Union Station blew off "with a roar like a boiler factory." In Boston, trees fell like leaves, and "slate roofing and shingles flew like confetti." For a two-minute interval, at Harvard University's Blue Hill Observatory, located 640 feet above sea level on Blue Hill just outside Boston, a wind gauge recorded a maximum average speed of 186 miles per hour. In Hartford, Connecticut, citizens fought a 36-foot flood through the night with sand bags. And everywhere there was desolation.

Long after the hurricane wore itself out the next day in Canada near Ottawa, its victims were still being identified. Thousands of people from New Jersey to Massachusetts were homeless; more than 650 were dead. Property losses exceeded $400,000,000. Without doubt, this "big blow" was the costliest natural disaster in our history.

Since 1938, weathermen have studied hurricanes closely with the aid of airplanes and radar. As a consequence, although nothing can be done to avert future storms, they will seldom arrive unheralded.

But unlike the hurricane, the tornado—the vacuum sweeper of the atmosphere and the monster of all winds—strikes so swiftly and unpredictably that it rarely gives fair warning.

In late spring and early summer, tornadoes visit the Mississippi and Missouri valleys more frequently than any other section of the country. They have, however, appeared in every state.

A tornado is most likely to be produced on a calm, hot, humid day. Large thunderstorms disturb the sky. Rain and hail pour downward. Suddenly from the depths of the clouds twists a sinuous cone, its point writhing downward toward the ground with an unearthly roar (Plate 11), and traveling from 25 to 40 miles per hour.

Like a giant magnet, it picks up everything in its path. Inside the funnel, updrafts of 100 to 200 miles per hour sweep the earth clean. The air pressure within the cone falls so low that houses and barns literally explode when the cone passes over them. This sucking action of the cone can lift trains from their tracks, and toss trucks playfully about. It may kidnap babies from their homes and then freakishly leave them unhurt in the branches of a tree a hundred yards away.

About the center of the storm play winds estimated to be 300 miles per hour. With a force 900 times as great as a 10-mile-an-hour breeze, they drive splinters into boards, iron shafts through cement walls, and straws into trees as easily as you can put a tack in a bulletin board.

Yet the violence of the tornado is limited in extent. Its diameter is about 1,000 feet. The usual southwest-to-northeast path of the tornado across the countryside is seldom longer than 300 miles, and often less than 10. At that, the storm can cause an unbelievable amount of damage. One of the most destructive tornadoes recorded struck Missouri, Illinois, and Indiana in 1925. It killed 689 persons and caused more than $16,000,000 property damage.

In the *Monthly Weather Review* of May, 1930, a Kansas farmer, Mr. Will Keller, describes how he was working in his fields near Greensburg one day when he glanced at the sky and saw three tornadoes hanging from an "umbrella-shaped cloud." He ran for his tornado shelter. From it Mr. Keller watched one of the three tornadoes rise above the ground and rush toward him.

At last the great shaggy end of the funnel hung directly overhead. Everything was as still as death. There was a strong gassy odor and it seemed that I could not breathe. There was a screaming, hissing sound coming directly from the end of the funnel. I looked up and to my astonishment I saw right up into the heart of the tornado. There was a circular opening in the center of the funnel, about 50 or 100 feet in diameter, and extending straight upward for a distance of at least one half mile, as best as I could judge under the circumstances. The walls of this opening were of rotating clouds and the whole was made brilliantly visible by constant flashes of lightning which zigzagged from side to side. Had it not been for the lightning I could not have seen the opening, not any distance up into it anyway.

Around the lower rim of the great vortex small tornadoes were constantly forming and breaking away. These looked like tails as they writhed their way around the end of the funnel. It was these that made the hissing sound. . . .

The tornado was not traveling at a great speed. I had plenty of time to get a good view of the whole thing, inside and out. . . . Its course was not in a straight line, but it zigzagged across the country, in a general northeasterly direction.

After it passed my place it again dipped and struck and demolished the house and barn of a farmer by the name of Evans. The Evans family, like ourselves, had been looking over their hailed-out wheat and saw the tornado coming. Not having time to reach their cellar they took refuge under a small bluff that faced to the leeward of the approaching tornado. They lay down flat on the ground and caught hold of some plum bushes which fortunately grew within their reach. As it was, they felt themselves lifted from the ground. Mr. Evans said that he could see the wreckage of his house, among it being the cook stove, going round and round over his head. The eldest child, a girl of 17, being the most exposed, had her clothing completely torn off. But none of the family were hurt. . . .

So if you should see a tornado waving its trunk at you, keep the advice of S. D. Flora in mind. Writing about tornadoes in the *Climate of Kansas,* he noted:

It is doubtful whether any building, except possibly a steel-reinforced concrete structure with numerous floors and inside partitions, can withstand the force of a fully developed tornado, and the latter type building is almost sure to have its windows demolished and its cornices blown off. A wooden building is reduced to kindling in almost the proverbial "twinkling of an eye" and the air is filled with flying debris. Brick buildings of the ordinary type in city blocks usually have their upper stories demolished and the lower floor filled with debris.

The best refuge when a tornado is seen approaching is to get underground. The outdoor caves, "cyclone cellars," so often constructed adjacent to Kansas farm homes for the storage of fruits and vegetables, furnish excellent protection for persons who reach them. The southwest corner of the basement of a frame house is usually safe, as tornadoes commonly move from the southwest and debris is ordinarily carried to the opposite side of the basement. The most advisable thing to do for a person caught in the open when one of these storms is close is to lie down, preferably in a low place. To remain erect is to invite injury by flying debris or being blown away. For a person caught in a business section of a city, the chances of escape become largely a matter of luck. In that case, probably the safest place is in a lower hallway of a substantial building, well away from outside doors or windows, and crouching against a partition which might support the weight of collapsing walls and floors.

It is entirely possible for a person in an automobile to outrun a tornado if the road ahead is clear, or even to drive at right angles to its path. A person on foot, if there is sufficient time, can often escape the narrow path of destruction by running at right angles to the direction from which the tornado is moving.

Do not expect the Weather Bureau to warn you too far in advance. The chance of forecasting the brief appearance and narrow path of a tornado is small.

All these winds—the tornado, hurricane, chinook, and mountain and valley breeze—are exceptional rather than everyday winds. Despite the flow of the "Prevailing Westerlies," they occur in particular places in certain seasons. Keep them in mind when consulting your newspaper weather map. But don't be disappointed if you fail to find one frequently!

The everyday winds, on the other hand, are easier to spot. In winter, these winds are speedier than in summer. They roar loudest in March and April. In April, 1934, the highest wind speed ever recorded in the United States—231 miles per hour—was observed on Mount Washington, New Hampshire, 6,288 feet above sea level.

In any month, however, wind speed is proportionate to the prevailing pressure system. The closer the lines of equal pressure (or isobars) are spaced on the weather map, the greater the speed of the wind.

In hilly country, the wind speed is lighter and gustier than over the plains, for it is retarded in detouring over the rolling landscape. Coming off a large body of water, the wind will be stronger than when flowing from one farmland to another. Chicago is a windy city because it lies in the path of a majority of frontal systems that cross the country. It is also windy because a large body of water lies to the northeast, and level plains to the west and south.

You need not live in Chicago to watch the wind go by. In any part of the country, you can get a rough estimate of tomorrow's atmospheric conditions by noting the direction and strength of the wind. You can judge direction and speed by observing the motions of flags and branches on trees, or the drift of smoke from a chimney (see Appendix 1 for guide to estimating wind speed). When the air is calm, flags hang limp; smoke rises vertically. A wind of 8 to 12 miles per hour stirs leaves and twigs constantly, and extends light flags. If it reaches 25 to 31 miles per hour, it will sway large tree branches, and whistle through telephone wires. By the time it passes 76 miles per hour, you have a wind of hurricane force on your hands. But, whether it is of hurricane or tornado force, or only an unspectacular calm; whether it is from the east, west, valley, or mountains; whether it is warm or cold; whether it is a breeze or a blow—each wind has a part in the passing parade of weather.

12

Forecasting the Weather: How to Be Your Own Ground Hog

> When the locks turn damp in the scalp house,
> Surely it will rain.
>
> —American Indians

THE ground hog is one of the best-known weather forecasters. As a prophet, he has an enviable reputation. Few people can claim that they are consulted religiously about the weather even once a year. The ground hog's reputation is not above praise, however, because it is based more upon the publicity given his forecasts than upon their accuracy. Although you may not rival him on the first count, you may surpass him on the second. For you can learn to predict major variations in the weather accurately.

To become proficient, don't watch for shadows on the ground. Look instead for changes that take place in the clouds, wind, and humidity, as well as those in pressure and temperature (see Appendix 2D). Then you'll be ready to come out on February 2nd, or on any other day in the year, and put the ground hog in his place.

The most obvious signposts of changing weather are the clouds (see Appendix 2B). Coming rain or snow is indicated when altostratus and stratocumulus clouds darken the horizon. Clouds like these, moving swiftly from the south, precede a cold front. If you live along the Atlantic coast, you can expect the cloudiness to clear up soon after the cold front flows by. Throughout the middle western states, stratocumulus clouds may persist for one or two days after a cold frontal passage.

When cirrus and cirrostratus drift overhead, followed

by thick, lowering altostratus, look for warmer weather. Since this sequence of clouds advances ahead of a warm front, you can count on a rise in temperature shortly.

Cirrus clouds by themselves may have little significance. In the southwest, they stand alone in the sky for days without meaning anything more than "still warmer."

Generally speaking, clouds moving from the south are heralds of precipitation, while those from the north signify clear weather. In all events, "the higher the clouds, the finer the weather"—for the moment. But "in dry weather, all signs fail."

A sudden rise in wind speed may also foretell a turn in the weather. Yet a significant increase in wind speed should not be confused with the daily rise and fall in the wind. In the middle of the afternoon, for instance, the average wind speed is highest; at daybreak, it is lowest. At any time of day, however, when the wind blows more breezily than usual, watch for fresh weather.

A quick shift in wind direction is likewise indicative of changing conditions. A shift in wind from east to south to west brings with it a greater variation in the weather than one from east to north to west. On the whole, winds from the east and south are foul-weather winds; northerly and westerly winds are fair-weather winds. So, "Do business with men when the wind is in the northwest." Remember the Spanish saying, though, that "When God wills, it rains with any wind."

In addition to the clouds and the wind, keep your eyes on the barometer. You can rely on it more in winter, when pressure systems are most pronounced. It is less reliable in summer, when much of the rain that falls comes from thunderstorms which barometer readings seldom anticipate.

But don't put too much faith in any one reading. When the dial points to "Stormy" or "Very Dry," this fact is not so important as the observation that the pressure has been steadily rising or falling.

During the day, there is a normal fluctuation in pressure. The highest pressure occurs about 10 o'clock in the morning, and the lowest about 4 o'clock in the afternoon. If, despite this fluctuation, the barometer is rapidly rising, you can usually anticipate clearing skies. On the other hand, when the pressure falls rapidly, look for a spell of bad weather (see Appendix 2A). Hence

> When the glass falls low
> Prepare for a blow;
> When it rises high,
> Let all your kites fly.

When the air pressure falls, you can often "smell" rain, for in the country the lowering pressure releases foul odors from swamplands where vegetation is decaying. The lessening pressure also causes old rheumatic pains to return, and

> A coming storm, your shooting corns presage,
> And aches will throb, your hollow tooth will rage.

In such wet weather, birds fly closer to the ground than usual to catch insects unable to rise far aloft with wet wings. Fish may swim near the surface of the water and bite more readily. The odor of flowers is most noticeable. Because the pink-eyed pimpernel closes its petals before rain falls, it is called the "peasant's weather glass." So, as Dr. Jenner wrote long ago, look for rain when

> The hollow winds begin to blow,
> The clouds look black, the glass is low,
> The soot falls down, the spaniels sleep,
> And spiders from their cobwebs creep.
>
> Last night the sun went pale to bed;
> The moon in halos hid her head.
> The boding shepherd heaves a sigh,
> For, lo!—a rainbow spans the sky. . . .
>
> Loud quack the ducks, the peacocks cry,
> And distant hills are looking nigh; . . .
> Low o'er the grass the swallow wings;
> The cricket, too, how sharp he sings!
>
> In fiery red the sun doth rise,
> Then wades through clouds to mount the skies;
> The walls are damp, the ditches smell
> For closed are pink-eyed pimpernels!

Your forecasts will be more accurate if they are based on observations of the temperature as well as of the barometer, the pains in your toes, and the quacking of ducks. You'll find that the temperature rises slightly before a storm. The most conspicuous rise takes place in winter. Furthermore, the greater the intensity of the storm, the greater will be the temperature change after it passes.

You'll note too that the average temperature variation from night to day is greater in winter than in summer. In either season, your forecast for temperature extremes should take into account the expected wind direction and speed, and the cloud cover and precipitation.

For example, when southerly winds are forecast to persist throughout the day, the temperature will normally rise. But if you also expect clouds which will shut off much of the sunlight, and rain which, upon evaporating, will lower the air temperature, your forecast for the maximum temperature should not be so high as it would be if the sky were clear.

When you supplement these observations with a glance at the weather maps published in the newspapers, you'll get a closer insight into the weather. You can quickly locate important cold and warm fronts, as well as occlusions. You can estimate their eastward movements, remembering that the stronger a front is, the more consistently it travels. Perhaps the best rule to use is that the systems will travel as far today as they did yesterday.

Nearly all strong fronts increase in intensity as they reach the Mississippi valley, particularly in winter. An allowance must be made for this fact when estimating eastward movement, for fronts slow down in speed as they build up in strength.

Moreover, the farther eastward a front travels, the nearer it comes to sources of moisture like the Atlantic Ocean and the Gulf of Mexico. Consequently, it is likely to have more and more cloudiness associated with it.

The intensity, orientation, and type of fronts that pass by have much to do with changes in the weather. By studying newspaper weather maps and also by watching the temperature, pressure, winds, and clouds, you'll be ready for what happens in the next day or two (see Appendix 2C and 2E).

For advice on the weather in the next week, consult the Weather Bureau five-day forecasts. These advisories are issued two days in advance of the forecast period. The preparation of the forecasts is complicated, for it involves the projection of surface and upper-air circulation patterns into the future. Nonetheless, the five-day forecasts are surprisingly reliable and are constantly increasing in accuracy.

The Weather Bureau also makes thirty-day forecasts on an experimental basis. But these forecasts aren't yet as dependable as its five-day ones.

To anticipate any longer weather trends, you must look elsewhere. Neither Weather Bureau forecasts nor evening cloudiness can point in September to a severe winter, or before Eastertime can foretell a dry summer. For this information, you must turn either to a backwoods weather prophet or a front-office climatologist.

According to the backwoods weather prophet, at Hallowe'en time you can expect a white winter when markings on the backs of caterpillars measure one inch or more, and when the number of skunks in the barn is unbearable. You can be sure of such a winter if cattail heads are longer and bushier than ever, and if wild geese have flown south earlier than usual.

For the front-office climatologist, skunks in the barn mean only "enter barn with caution," and not "two feet of snow by Christmas." He prefers to rely instead on past weather data and to speculate on their recurrence.

His records, for instance, show that precipitation falls over the country in definite patterns in different seasons of the year. Moreover, these patterns recur each year in much the same fashion (Plate 10). By studying these patterns and applying them to long-range forecasting, you can anticipate dry weather in California during the summer and rain in the winter with greater certainty than if you relied only on cattails and geese.

These climatological data demonstrate that two to four months are rainless in summertime in the valleys of California and along the Sierra Nevada Mountains. Extensive irrigation is necessary to grow oranges and grapefruit. Grass and flowers must be watered continually. This state of the climate is a translation of the fact that the great area of high pressure, the Pacific High, dominates the region in summer. When the High moves southward in winter and leaves the coastline unprotected, you can count on widespread precipitation. Occlusions (which during the summer are kept from entering this area by the Pacific High) repeatedly move inland and leave both rain and snow in their wake. Sacramento, California, which averages slightly more than 18 inches of rainfall a year, is located in this region.

If you live in the states of Arizona, New Mexico, or in the southern parts of Utah, Nevada, and California, you can also expect a period of rainy weather in winter due to the passage of numerous occlusions over those states. July and August are "cloudburst" months. During this time, heavy thundershowers scatter across the mountains when moist air flows upslope from the Gulf of California or the Gulf of Mexico. June, however, is dry. Albuquerque, New Mexico, in the eastern part of this area, has an average of 8 inches of precipitation per year (average precipitation or rainfall figures include the water from the yearly snowfall). The state of Nevada, which averages less than 9 inches of precipitation each year, is the driest in the United States—from the point of view of the weather.

Within this region, a large heat or "thermal" low-pressure area develops during the summer. The air in the "thermal" low is hot, dry, and light. Mountains surround the area and cut off much of the flow of moist air from the Pacific Ocean and the gulfs of Mexico and California. As a result, there is little cloudiness. Since skies are clear and since the air has no large body of water nearby to keep it cool, temperatures inland often rise to 110 to 115 degrees in the afternoon.

Death Valley, California, is located in this "thermal" low at a spot where all the air that passes over it must flow downhill from 3,000 or 4,000 feet to below sea level. This air is heated about 5½ degrees for every 1,000 feet it falls. Already warm, it becomes deathly hot. It is little wonder, therefore, that the second highest temperature in the world, 134 degrees, was recorded in Death Valley.

For northern Utah and Nevada, Washington, Oregon, and Idaho, a long-range climatological forecast of heavy rain and snow during December and January is safe. Spring and autumn are likewise well-watered seasons. This rainfall pattern shows that the occlusions which form in the maritime polar air over the Pacific Ocean, and then pass through these states all year, are strongest in winter. In this section, Boise, Idaho, averages about 13 inches of rainfall a year.

Summer is umbrella-time in the triangularly shaped region bounded by a line running between Great Falls, Montana, Sault Ste. Marie, Michigan, and San Antonio, Texas. The water comes mainly from thunderstorms created in moist maritime tropical air flowing northward from the Gulf of Mexico. Winter is relatively dry—and decidedly cold. Omaha, Nebraska, at the center of this region, normally receives 28 inches of precipitation a year, 1 inch less than the over-all average for the United States.

Look for spring as the wettest part of the year in Tennessee, Kentucky, western North Carolina, northern Georgia, and Alabama. Count on autumn to be the driest season. At that time, the Atlantic High pressure area, which covers the region during summer as well as autumn, is strongest. Nashville, Tennessee, has an average of 46 inches of precipitation each year.

A guess of an unvarying rainfall pattern from season to season is usually right for the states of New York, New Jersey, New England, and the St. Lawrence valley. Frontal systems move over this district so regularly that the normal rainfall is almost the same from month to month. New York City annually totals about 43 inches of rainfall.

For the rest of the Middle Western and eastern states, expect summer to be the wettest season. During the summer, thundershowers produced in maritime tropical air, coupled with precipitation from warm fronts roaming the country, bring abundant rain. In addition, cold, warm, and occluded fronts pass through the area each season. Furthermore, these states are closest to the Gulf of Mexico, so they average more rain each year than any other region. Louisiana is the wettest state in the country with its average yearly rainfall of 55 inches; Chicago, Illinois, has 33 inches.

These average rainfall figures do not necessarily coincide with those for cloudiness. Just because there is little rainfall, you cannot count on an absence of clouds. In fact, clouds may be plentiful even when rain is scarce; in the summertime California has almost no rain despite the fact that stratus clouds hug the coastline much of the season.

Throughout the country, the average cloudiness for the year is greatest in the state of Washington. It is lowest in Nevada, southern California, and Arizona. Clouds dot the sky about one-half of the time in the Middle West, southeast, and northeast.

All these climatological data, consequently, can give you a picture of what the *average* weather conditions will probably be next month or next year. But since the weather often doesn't live up to its past standards, you can't be absolutely certain that next April will be rainy. You can be certain of only one thing—that the weather will keep on following its familiar habits.

Without doubt, five hundred years from today the clouds will still continue to decorate the sky with their intricate patterns. Despite the hydrogen bomb, the fog will continue to steal through the back yard at night. The winds will still vie with each other in speed, and the fronts in strength. Then, as now, the weather will be variable. Then, as now, there will be disagreeable days and there will be invigorating days. It will rain, as it has frequently done, on the Fourth of July, and it will refuse to snow for Christmas. And all the threats in the world will be to no avail.

So the best you can do about tomorrow's weather is to forecast it today. You can do this, with gratifying results, by studying the sky and wind, the barometer and the thermometer, and the weather maps, forecasts, and data. By depending upon these atmospheric signals instead of the nearest ground hog, you'll be prepared to follow the changing weather in a changing world.

Appendices

1

How to Estimate the Wind Speed

Beaufort Number	Weather Maps	Speed Miles per Hour	General Description	Specifications
0	○	less than 1	calm	Smoke rises vertically.
1		1 to 3	light air	Wind direction shown by drift of smoke.
2		4 to 7	slight breeze	Wind felt on face; leaves rustle.
3		8 to 12	gentle breeze	Leaves and twigs in constant motion; wind extends light flags.
4		13 to 18	moderate breeze	Dust, loose paper, and small branches are moved.
5		19 to 24	fresh breeze	Small trees in leaf begin to sway.
6		25 to 31	strong breeze	Large branches in motion; whistling in telegraph wires.
7		32 to 38	moderate gale	Whole trees in motion.
8		39 to 46	fresh gale	Twigs break off trees; walking is impeded.
9		47 to 54	strong gale	Slight damage to houses; chimney pots blown off.
10		55 to 63	whole gale	Trees uprooted; considerable damage to houses.
11		64 to 75	storm	Rarely experienced; widespread damage.
12		above 75	hurricane	Excessive damage.

2

Forecasting Guides for the Layman

A. Wind and Barometer Indications

Wind direction (Note: A NW wind blows from the northwest.)	Sea-level barometric pressure in inches	Character of weather indicated
SW to NW	30.10 to 30.20 and steady	Fair, with little temperature change, for 1 to 2 days.
SW to NW	30.10 to 30.20 and rising rapidly	Fair, followed within 2 days by warmer and rain.
SW to NW	30.20 and above and stationary	Continued fair with no marked temperature change.
SW to NW	30.20 and above and falling slowly	Fair with slowly rising temperature for 2 days.
S to SE	30.10 to 30.20 and falling slowly	Rain within 24 hours.
S to SE	30.10 to 30.20 and falling rapidly	Wind increasing in force, with rain within 12 to 24 hours.
SE to NE	30.10 to 30.20 and falling slowly	Rain in 12 to 18 hours.
SE to NE	30.10 to 30.20 and falling rapidly	Increasing wind and rain within 12 hours.
E to NE	30.10 and above and falling slowly	In summer, with light winds, rain may not fall for several days. In winter, rain within 24 hours.
E to NE	30.10 and above and falling rapidly	In summer, rain probably within 12 to 24 hours. In winter, rain or snow, with increasing winds.

Wind direction (Note: A NW wind blows from the northwest.)	Sea-level barometric pressure in inches	Character of weather indicated
SE to NE	30.00 or below and falling slowly	Rain will continue 1 to 2 days.
SE to NE	30.00 or below and falling rapidly	Rain with high winds, followed within 36 hours by clearing, and in winter by colder.
S to SW	30.00 or below and rising slowly	Clearing within a few hours and fair for several days.
S to E	29.80 or below and falling rapidly	Severe storm soon, followed within 24 hours by clearing, and in winter by colder.
E to N	29.80 or below and falling rapidly	Severe northeast gale and heavy precipitation. In winter, heavy snow followed by a cold wave.
Going to W	29.80 or below and rising rapidly	Clearing and colder.

NOTE: This table, prepared by the U.S. Weather Bureau, is a general summary of observations taken all over the country. It is, therefore, an average, and will not apply to your back yard without some alteration. The pressure readings in particular differ in different localities. This table, however, will help you to correlate wind and barometric tendencies with changes in the coming weather.

A *rapid* rise or fall of pressure would be one equal to or greater than .045 to .09 inches (2 to 3 millibars) in three hours. A *slow* rise or fall of pressure would be less than this amount, depending again on the locality and circumstances.

B. Clouds and Their Meaning

High Clouds (average lower level above 20,000 feet)

Name and Appearance	Composed of Water Droplets or Ice Crystals	Accompanying Precipitation	What They Say About the Future Weather
Cirrus are detached, delicate, fibrous, silky, without shading, but with ragged, indefinite edges. They are often colored brilliantly at sunrise or sunset.	Ice	None	Fair weather if they do not thicken. But rain or snow in 12 to 24 hours if they are followed by thicker, lower clouds preceding a warm front.
Cirrocumulus come in layers or patches of small, white flakes without shading, and are often said to form a "mackerel sky." They always appear in the company of some other type of cirrus cloud.	Ice	None	When followed by lower and thicker clouds, look for rain and warmer temperatures.
Cirrostratus cover the sky with a thin, whitish veil. When they pass in front of the sun or moon, a *large* halo-like circle is produced with red on the inside and blue on the outside of the colored ring.	Ice	None	Fair weather if they degenerate into cirrocumulus. If they occur early in evening, they may clear up before morning. Yet they often mean rain and higher temperatures to follow, because they precede warm fronts.

Middle Clouds (average lower level 6,500 feet; average upper level 20,000 feet)

Name and Appearance	Composed of Water Droplets or Ice Crystals	Accompanying Precipitation	What They Say About the Future Weather
Altocumulus occur in small, isolated patches, parallel bands, or in a layer, composed of flat-	Usually water	Occasional light rain or snow from thick clouds. Often trail-	If corona becomes smaller with time, look for rain. Clouds often merge into rainy altostratus

Name and Appearance	Composed of Water Droplets or Ice Crystals	Accompanying Precipitation	What They Say About the Future Weather
tened globular masses. When they pass in front of the sun or moon, a corona forms *small*, rainbow-like rings with red on the outside and blue on the inside.		ing precipitation called *virga*.	which precede warm fronts. Clouds also precede cold fronts.
Altostratus produce a gray or bluish fibrous veil without a halo and of variable thickness. The sun appears as though it were shining through ground glass and does not cast shadows.	Usually water, but upper portions may turn to ice.	Steady rain or snow when clouds have icy tops.	Mean bad weather when they're increasingly dark to the west. Precede both warm and cold fronts.
Low Clouds (average lower level close to the surface; average upper level 6,500 feet)			
Stratocumulus come as a continuous sheet or patches of clouds composed of rounded masses or rolls, which are soft and gray, with darker parts.	Water	Light drizzle or snow flurries infrequently.	Since these clouds often form in the vicinity of thunderstorms as well as warm and cold fronts, be on the watch for changing weather.
Stratus are uniform clouds of indefinite shape which become *fog* if they rest on the ground. They give the sky a hazy appearance. A stratus cloud is often only 50 to 100 feet thick, so that clear sky can be seen through it.	Usually water	Light drizzle or snow flurries infrequently.	Often indicative of fair weather to follow. Nighttime stratus often clear up after the sun rises. Watch for fog in the mountains when stratus lie in valleys.

Name and Appearance	Composed of Water Droplets or Ice Crystals	Accompanying Precipitation	What They Say About the Future Weather
Nimbostratus are "rain clouds" like stratus but thicker. They are of a dark, uniformly gray color. Their bases look ragged and wet.	Ice and water	Steady rain or snow.	Arrive slightly ahead of warm fronts and can bring long, steady rain.
Clouds with Vertical Development (average lower level 1,600 feet; average upper level that of cirrus)			
Cumulus are dense clouds with vertical development which never overcast the sky, with dome-shaped upper surfaces and horizontal bases.	Water	None	They're "fair weather" clouds when they stand alone in the sky. They often appear in the cold air behind a cold front, indicating cool but fair weather ahead. At night, they often melt into stratocumulus or altocumulus.
Heavy and swelling cumulus boil upwards with great vertical development but they have no "anvil" top of cirrus clouds.	Water	None	When they form early on a summer morning, they mean thunderstorms and rain before the end of the day. When they form after the noon hour, rain is less likely. They also may precede a warm front and indicate higher temperatures in the offing.
Cumulonimbus are heavy, cauliflower-shaped masses of clouds with great vertical development whose tops are	Water and ice	Heavy rain or snow, with sometimes hail.	May precede either strong warm or cold fronts, and cause static heard on radios.

composed of an "anvil" of cirrus clouds. From the ground, the ragged bases of cumulonimbus look dark and menacing. If rain or snow is falling, a cloud base may appear to extend to the ground.

C. Weather within an Air Mass

	For an Air Mass That Is Colder than the Land over Which It Passes	For an Air Mass That Is Warmer than the Land over Which It Passes
Clouds:	cumulus, swelling cumulus	stratus, stratocumulus, fog, possibly nimbostratus
Height of Cloud Bases:	well above the ground	close to or on the ground
Visibility:	good	poor; often haze, smoke, or smog
Is Air Smooth or Bumpy for Flying?	bumpy close to the ground	smooth
Precipitation:	thundershowers with possibly hail rainshowers or snowshowers	drizzle or steady rain or snow

Note: This is a very generalized picture and often won't apply in specific cases. The tendency for air masses to behave in this manner, however, is well recognized, although the presence of lakes or mountains, the time of day or night, and the general weather trend will modify this picture.

Examples of air masses that are colder than the land over which they pass are continental polar air in winter moving southward from Canada, maritime tropical air in summer moving northward from the Gulf of Mexico, and maritime polar air in summer moving from the Pacific over the west coast.

Examples of air masses warmer than the land over which they pass are maritime tropical air in winter moving northward from the Gulf of Mexico, continental polar air moving southward over the Great Lakes in summer, and maritime polarair moving over the west coast in winter.

Visibility is defined as the maximum distance at which large objects can be distinguished with the naked eye. On days when visibility is exceptional within an air mass, the air temperature will be cooler than normal, indicating the presence of an air mass colder than the land over which it passes.

D. General Rules for Guessing
(To be used with tons of salt)

Look for the weather to remain fair when:

summer fog breaks up before noon.
the bases of clouds along the mountains increase in height.
clouds tend to decrease in number.
the wind blows gently from the directions of west to northwest.
the temperature is normal for the time of year.
the barometer is steady or slowly rising.
the setting sun looks like a ball of fire and the sky is clear.
the moon shines brightly and the wind is light.
there is heavy dew or frost at night.

Look for a change in the weather for the worse when:

cirrus clouds change into cirrostratus, and cloudiness is thickening and darkening to the west or southwest.
quickly moving clouds increase in number and lower in elevation.
clouds move in various directions at different elevations.
clouds move from the south and the southerly wind increases in speed.
altocumulus or altostratus clouds darken the western horizon and the barometer falls rapidly.
the sky is clear at sunset, the wind speed light, and the air moist (look for fog).
there is sustained flow of moist, warm air northward from the south over colder ground (look for fog).
the wind shifts to the south or east; the largest change comes when the wind shifts from north to east to south.
the wind blows strongly in the morning.
the temperature rises conspicuously in the winter.
the temperature is far above or below normal for the time of year.
the barometer falls steadily.
bunions or teeth ache as a result of a rapid fall in pressure.
there is a hard rainfall at night.
a warm front approaches.
a cold front approaches.

Look for clearing weather when:

cloud bases rise farther aloft.
a cloud-filled sky shows signs of clearing up.
the wind shifts to a westerly direction; the largest change comes with a shift from east to south to west.
the barometer rises rapidly.
a strong cold front has passed 3 to 6 hours ago.

Look for rain or snow: (the time intervals here are only approximate; the faster a storm moves, the shorter the time intervals become)

when a cold, warm, or occluded front approaches.
18 to 36 hours after the first cirrus clouds are spotted in the sky (provided they thicken and give way to lower clouds).
12 to 24 hours after cirrus clouds thicken into cirrostratus and a corona is seen around the sun or moon.
within 6 hours when the morning temperature is high, the air is moist and sticky, and a few cumulus clouds ride the sky.
within 1 hour in the afternoon when there's static on the radio, swelling cumulus clouds overhead, and a dark sky to the southwest.

Look for the temperature to fall when:

the wind continues to blow from the north or northwest.
the wind shifts into the north or northwest.
the wind is light and the evening sky is clear.
the pressure rises (in winter).
a cold front has passed.

Look for the temperature to rise when:

the sky is filled with clouds at night and there's a moderate southerly wind.
the sky is clear all day and the wind is from the south.
the wind shifts from the northwest to the south.
a warm front has passed.

E. Warm Fronts, Cold Fronts, and Weather

I. A Warm Front

	Ahead of the warm front	At the moment the warm front arrives	Behind the warm front
Wind:	Increases in speed; direction often easterly	Changes direction to southeasterly	Changes direction into the southwest

	Ahead of the warm front	At the moment the warm front arrives	Behind the warm front
Barometer:	Falls—the faster the fall, the sooner the surface front will arrive	Reaches its lowest reading as "trough" of warm front arrives	Often only slight rise
Temperature:	Slight rise	Slow (but not sharp) rise	Rise—amount of rise depends upon the temperature difference between cold air ahead of front and warm air behind it
Humidity:	Increases slowly	More rapid increase to nearly 100 per cent	Slight increase possible
Visibility:	Continually worsening	Poor, sometimes thick mist or fog	Slight improvement possible
Clouds:	Cirrus are sighted first overhead, and then are followed by cirrostratus or altocumulus, and finally by stratus or nimbostratus	Nimbostratus or (occasionally) cumulonimbus	Occasionally scattered nimbostratus, stratus, or stratocumulus. But generally marked clearing of bad weather
Weather:	Drizzle from altostratus, then steady rain or snow from nimbostratus	Drizzle, rain snow, or showers	Occasional rain or snow possible, but most often clearing weather

NOTE: This is the usual sequence of weather elements which accompanies the movement of a warm front. It is a generalized one, however, and often will not apply to specific cases.

The rapidity with which these weather elements appear and disappear depends, of course, upon the speed at which the warm front moves. The faster it travels, the more rapidly the weather elements appear and disappear.

The magnitude of their changes and the length of time they persist depends again on the speed of the warm front, but also upon the season, the time of day or night, the type of air masses that are separated by the warm front and the temperature difference between them, and the general weather situation.

II. A Cold Front

	Ahead of the cold front	At the moment the cold front arrives	Behind the cold front
Wind:	Increases in speed; direction often easterly	Rapid change of direction into the west with gusts and squalls	Steady but strong from the west to northwest
Barometer:	Rapid and continuous fall—the faster the fall, the sooner the surface front will arrive	Reaches its lowest reading as "trough" of cold front arrives	Rapid rise
Temperature:	Continuous rise	Little change; then	Sudden drop—amount of fall depends upon the temperature difference between warm air ahead of front and cold air behind it
Humidity:	Steady	Decrease	Well-defined fall (except during showers)
Visibility:	Poor, with possible mist or fog (with low wind speeds)	Poor to good, depending upon intensity of precipitation accompanying the arrival of the front	Excellent

	Ahead of the cold front	At the moment the cold front arrives	Behind the cold front
Clouds:	Altocumulus or altostratus give way to stratus, stratocumulus and cumulonimbus or (occasionally) nimbostratus	Cumulonimbus with low stratus; occasionally nimbostratus	Altocumulus and cumulus or stratocumulus, but rapid decrease in amount of clouds and marked rise in the height of their bases
Weather:	Fog, rain, snow, or thundershowers increasing in intensity to—	Heavy rain or snow; sometimes beginning of winter "blizzard"	Occasional scattered showers, but generally clearing weather

NOTE: This is the usual sequence of weather elements which accompanies the movement of a cold front. It is a generalized one, however, and often will not apply to specific cases.

The rapidity with which these weather elements appear and disappear depends, of course, upon the speed at which the cold front travels. The faster it travels, the more rapidly the weather elements appear and disappear.

The magnitude of their changes and the length of time they persist depends again on the speed of the cold front, but also upon the season, the time of day or night, the type of air masses that are separated by the cold front and the temperature difference between them, and the general weather situation.

3

Books, Facts, and Figures for the Layman

To subscribe to the daily weather maps published by the Weather Bureau, United States Department of Commerce, address money order or check payable to the Treasurer of the United States, Superintendent of Documents, Government Printing Office, Washington 25, D.C. The maps are prepared daily, including Sundays and holidays. The price is 60c a month or $7.20 a year.

Other publications obtainable at the Government Printing office:

Climate and Man, Yearbook of Agriculture, 1941, 1248 pp. Cloth $1.75. (Now out of print.)
A well-illustrated book which discusses world climate since the beginning and its effect upon man. Included in the book are interesting articles by weather specialists as well as material about the climates of each of the forty-eight states.

Codes for Cloud Forms and States of the Sky, 1938, 100 pp. 30c.
A profusely illustrated dictionary of clouds which is indispensable to every amateur weatherman.

Meteorology for Pilots, 2d ed., 1943, 246 pp. $1.25.
This book, although written for pilots, is an excellent one for amateur weathermen who want to investigate the theoretical side of meteorology, for it is well-diagramed. However, it is not easy reading for the beginner who isn't interested in equations and theory.

Realm of Flight, 1947, 41 pp. 60c.
Prepared by the Civil Aeronautics Administration to give the private pilot information essential to the safe operation of his aircraft, this pamphlet contains much material of interest to the amateur weatherman. The colored drawings are easy to understand.

Weather, Astronomy, and Meteorology, Price List 48, 38th Edition, Sept., 1948.
A list of all publications for sale by the Government Printing Office which relate to weather, astronomy, or meteorology.

Weather Glossary, 1946, 299 pp. 65c.
A glossary containing weather words from "ablation" to "zonda," and useful to the amateur weatherman who takes his hobby seriously.

Other articles and magazines of interest to the layman:

House Beautiful magazine began a series of articles on "Climate Control Studies" in its issue of October, 1949. These articles are easy to follow and are of interest to every home owner. If you're planning to build a home, by all means look up these studies.

The National Geographic Magazine and *Harper's Magazine* occasionally publish weatherwise articles.

Weatherwise, The Magazine About Weather, is published bimonthly by the American Meteorological Society, 3 Joy Street, Boston 8, Mass. The subscription rate for this popular magazine is $3.00 per year. Make checks or money orders payable to the American Meteorological Society.

Several books about the weather which aren't too technical:

Blair, T. A. *Weather Elements*, Prentice-Hall, 1948, 373 pp.
A fine textbook which discusses all phases of the weather. The tough parts can be skipped without losing too much of the continuity.

Conant, J. B. *On Understanding Science*, Yale University Press, 1948; The New American Library, 1951, 144 pp. A fascinating book which explains how scientific techniques are developed, and has a good deal of material relating to the weather.

Flora, S. D. *Climate of Kansas*, Report of the Kansas State Board of Agriculture, Vol. LXVII, No. 285, June 1948, 320 pp.
A thorough study of the Kansas climate, with many charts and pictures. If you farm or do business in Kansas, it's a handy book to have. If you're interested in the climates of the United States, get it.

Grant, H. D. *Cloud and Weather Atlas*, Coward-McCann, 1944, 294 pp.
A book for amateur weathermen who want something more complete than the Weather Bureau's *Codes for Cloud Forms and States of the Sky*. The cloud pictures in this book are beautiful and the discussion that accompanies them is entertaining.

Inwards, R. *Weather Lore*, 4th edition edited by Hawke, London, 1950, 251 pp. A book devoted to a vast collection of sayings and proverbs on weather lore. A "must" for people who are more interested in woolly bears than barometers.

Kendrew, W. G. *The Climates of the Continents*, Oxford, 1942, 473 pp.
If you want to know the month of maximum average rainfall in Italy, this is the book for you. It discusses clearly and simply

the climates of all the continents and is fascinating if you can skim over the tons of figures.

Kraght, P. E. *Meteorology for Ship and Aircraft Operation*, Cornell Maritime Press, 1943, 373 pp.
A good textbook for the amateur weatherman as well as for pilots of planes and boats. The many illustrations are excellent; the theory can be overlooked in favor of the more easily understandable drawings.

Leet, L. D. *Causes of Catastrophe*, McGraw-Hill, 1948, 232 pp.
If you're wild about hurricanes, earthquakes, volcanoes, tidal waves, or geophysical fables, you'll like this book. If you don't know anything about them, you'll find this well-written book full of amazing but true facts which can be used to awe your neighbors. For some reason, tornadoes are not discussed.

Nordhoff and Hall. *The Hurricane*, Little, Brown & Co., 1936, 257 pp.
A well-known novel about what a hurricane did to a South Sea island and its inhabitants.

Stewart, G. B. *Fire*, Random House, 1948, 336 pp.
Storm, Modern Library, 1947, 349 pp.
In both of these novels, the weather is one of the main characters. *Storm* is a fine book for the layman who wants to get an over-all picture of how the weather moves.

Tannehill, I. R. *Hurricanes*, Princeton University Press, 1942, 265 pp.
An excellent discussion about all phases of hurricanes. Technical in parts, this is a textbook for lovers of big blows.

Visher, S. S. *Climate of Indiana*, Indiana University Publications, 1944, 511 pp.
The first complete book ever written about the climate of one state. The material is not so attractively presented as it is in the *Climate of Kansas*, but the data are just as valuable.

Wenstrom, W. H. *Weather and the Ocean of Air*, Houghton Mifflin, 1942, 484 pp.
A lengthy book about weather and its relation to flying as well as to the man in the street.

More advanced books for the serious students of weather include:

Brooks, C. E. P. *Climate Through the Ages*, Ernest Penn Ltd., London, 1950, 395 pp.

Compendium of Meteorology, American Meteorological Society, 1951, 1315 pp.

Conrad and Pollak. *Methods in Climatology*, Harvard University Press, 1950, 459 pp.

Geiger, R. *The Climate Near the Ground*, Harvard University Press, 1950, 482 pp.

Haurwitz, B. *Dynamic Meteorology*, McGraw-Hill, 1941, 365 pp.

Haynes, B. C. *Techniques of Observing the Weather*, Wiley & Sons, 1947, 272 pp.

Humphreys, W. J. *Physics of the Air*, McGraw-Hill, 1940, 676 pp.
Landsberg, H. *Physical Climatology*, Penn. State College, 1943, 283 pp.
Neuberger, H. *Introduction to Physical Meteorology*, Penn. State College, 1951, 271 pp.
Petterssen, S. *Weather Analysis and Forecasting*, McGraw-Hill, 1940, 505 pp.

Weather Extremes for the United States:

Temperature
Lowest: -66 degrees at Riverside Ranger Station, Yellowstone Park, Feb. 9, 1933.
Highest: 134 degrees at Greenland Ranch, Death Valley, Calif., July 10, 1913.

Pressure
Lowest: 26.35 inches (892 millibars) at the Florida Keys, Sept. 2, 1935.
Normal Sea-Level: 29.92 inches (1013 millibars).
Highest: 31.5 inches (1067 millibars).

Wind
Highest: Gust of 231 miles per hour at Mt. Washington, N. H., April 1934.

Precipitation (includes both rain and snow, on the basis that 10 inches of snow are equal in water content to approximately 1 inch of rain).
Driest State: Nevada, with a yearly average of 8.8 inches.
Driest Spot: Greenland Ranch, Death Valley, Calif., with an annual average of 1.35 inches.
Average for U.S.: About 29 inches.
Wettest State: Louisiana, with a yearly average of about 55 inches.
Wettest Spot: Wynoochee Oxbow, Washington, with a 13-year average annual rainfall of 150.7 inches.
Largest rainfall in 24 hours: 23.22 inches at New Smyrna, Fla. Oct. 10-11, 1924.
Largest snowfall: in 24 hours: 60 inches at Giant Forest, Calif.
during one season: 884 inches at Tamarack, Calif., during season of 1906-7.
average seasonal: 451 inches at Tamarack, Calif.

Hail
Largest single hailstone: 1½ pounds, which fell at Potter, Neb., July 6, 1928.

Comparison of Fahrenheit with Centigrade scale:

Freezing 32 degrees Fahrenheit (F)
0 degrees Centigrade (C)

Boiling 212 degrees F
100 degrees C

To convert F to C, subtract 32 and multiply by 5/9.

Thus 212 degrees F = $(212-32) \times 5/9 = 100$ degrees C
176 degrees F = $(176-32) \times 5/9 = 80$ degrees C

To convert C to F, multiply by 9/5 and add 32.

Thus 0 degrees C = $(0 \times 9/5) + 32 = 32$ degrees F
80 degrees C = $(80 \times 9/5) + 32 = 176$ degrees F

Index

"You have a thing for cranky, ill-tempered men?"

"Apparently so," she said with a deliberate air of resignation.

Ryan's lips curved into a full-fledged grin then. "Lucky me."

She grinned back at him. "Try to remember that."

"Oh, I imagine you're going to give me plenty of occasions to question it," he said.

She nodded. "It is my mission, remember?"

"Maggie—"

She touched a finger to his lips to silence him. "Just accept it. I'm here to stay."

"But why?" he asked, obviously bewildered.

"It's that cranky, ill-tempered man thing," she reminded him. "I'm a sucker for a challenge." She hooked her hand around his neck and drew his head down till she could kiss him. "And it doesn't hurt that you're a great kisser!"

Dear Reader,

Have you ever been so excited after reading a book that you're bursting to talk about it with others? That's exactly how I feel after reading many of the superb stories that the talented authors from Silhouette Special Edition deliver time and again. And I'm delighted to tell you about Readers' Ring, our exciting new book club. These books are designed to help you get others together to discuss the brilliant and involving romance novels you come back for month after month.

Bestselling author Sherryl Woods launches the promotion with *Ryan's Place* (#1489), in which the oldest son of THE DEVANEYS learns that he was abandoned by his parents and separated from his brothers—a shocking discovery that only a truly strong woman could help him get through! Be sure to check out the discussion questions at the end of the novel to help jump-start reading group discussions.

Also, don't miss the other five keepers we're offering this month: *Willow in Bloom* by Victoria Pade (#1490); *Big Sky Cowboy* by Jennifer Mikels (#1491); *Mac's Bedside Manner* by Marie Ferrarella (#1492); *Hers To Protect* by Penny Richards (#1493); and *The Come-Back Cowboy* by Jodi O'Donnell (#1494).

Please send me your comments about the Readers' Ring and what you like or dislike about what you're seeing in the line.

Happy reading!

Karen Taylor Richman,
Senior Editor

Please address questions and book requests to:
Silhouette Reader Service
U.S.: 3010 Walden Ave., P.O. Box 1325, Buffalo, NY 14269
Canadian: P.O. Box 609, Fort Erie, Ont. L2A 5X3